# Interpreting Spectra of Organic Molecules

**A Series of Books in Organic Chemistry**

ROBERT G. BERGMAN, Editor

# Interpreting Spectra of Organic Molecules

## THOMAS N. SORRELL

University of North Carolina at Chapel Hill

**UNIVERSITY SCIENCE BOOKS**
**Mill Valley, California**

University Science Books
20 Edgehill Road
Mill Valley, CA 94941

Production manager: Mary Miller
Copy editor: Aidan Kelly
Cover and text designer: Robert Ishi
Technical illustrator: John Foster
Typesetter: Asco Trade Typesetting Limited
Printer and binder: Maple-Vail Book Manufacturing Group

Library of Congress Catalog Card Number: 88-050538

ISBN 0-935702-59-8

Printed in the United States of America
10  9  8  7  6  5  4  3  2  1

*To Liz and Courtney*

# Contents

# Preface

Spectroscopy has become an integral part of organic chemistry, and most introductory organic chemistry textbooks present at least the fundamental aspects of this important subject. In addition, there are several texts on the market devoted specifically to correlating spectra and structure of organic molecules. In my own teaching, I have encountered two problems with the standard texts. First, they develop the subject of spectroscopy by introducing the data for each functional group separately. Anyone who has been involved with research knows that such an approach is limited, especially when working with unknown compounds. Second, many students taking chemistry have only a peripheral interest in the subject and do not want to be burdened with a lot of facts and information that will be of seemingly little use later in life.

At the University of North Carolina at Chapel Hill, we introduce students to the practice of interpreting IR, NMR, $^{13}$C NMR, and mass spectra in our second semester organic laboratory course in which they are working with unknown compounds (qualitative organic analysis). This is accomplished with interactive microcomputer software developed at UNC, with funding from the ACIS group at IBM. It was during the development of the software that I became aware of the need for an introductory-level text that emphasizes the strategies for interpreting spectra, and I began work on this book.

My aim has been to create a text to supplement either a lecture or laboratory course in organic chemistry. I have approached the subject by trying to outline general and useful approaches for interpreting spectra of simple compounds. The idea is to provide students who continue to study chemistry with a solid foundation for a more advanced course later. Therefore, I have touched only very lightly on some traditional topics like theory and instrumental design and operation. I have tried to keep the textual material to a minimum and have included as much data as possible in figures and tables so that they will be easily accessible for other courses or for research.

I have not included spectroscopy problems within the text since the aforementioned software is available. The software includes data (IR, NMR, $^{13}$C NMR, and mass spectra) for 20 compounds and questions (with answers) for 12 of those compounds. The questions focus the user's attention on the appropriate features of each spectrum that are necessary to deduce the structure for each unknown. Sufficient help screens in the form of tables and figures are included to assist learning the factual material that is included in this text.

Many people have contributed to help make this book a reality. I appreciate the many students at UNC who have given me useful comments about the software, many of which have helped me focus on what I needed to include in this text. I especially thank Martha Garrity for recording many of the spectra that appear in the book and for her valuable comments on the entire manuscript; and Professors Bob Bergman at UC Berkeley and Bob Hanson at St. Olaf's College, who offered many helpful suggestions and good critical reviews during the draft stages of the manuscript. Finally, I thank my wife Liz for her encouragement during the entire project.

*Thomas N. Sorrell*
*Chapel Hill, 1988*

# Acknowledgments

The author is grateful to the following publishers for permission to reprint certain material in this book.

Figure 2.23, page 38: From *Modern Methods of Chemical Analysis*, first edition, by R. L. Pecsok and L. D. Shields, Figure 11-8 ("Spectral Patterns for Substituted Benzenes in the 5 to 6$\mu$ Region"). Copyright © 1968 by John Wiley & Sons, Inc. Reprinted by permission of John Wiley & Sons, Inc.

Figure 2.32, pages 48–49: From *Modern Methods of Chemical Analysis*, first edition, by R. L. Pecsok and L. D. Shields, Figure 11-7 ("Colthup Chart"). Copyright © 1968 by John Wiley & Sons, Inc. Reprinted by permission of John Wiley & Sons, Inc.

Figure 3.21, page 77: From *Proton and Carbon-13 NMR Spectroscopy: An Integrated Approach* by R. J. Abraham and P. Loftus, Figure 2.1. Copyright © 1983 by Wiley Heyden, Ltd. Reprinted by permission of John Wiley & Sons, Inc.

Table 3.3, page 79: From *Modern Methods of Chemical Analysis*, first edition, by R. L. Pecsok and L. D. Shields, Table 12-3 ("Approximate Chemical Shift of Protons"). Copyright © 1968 by John Wiley & Sons, Inc. Reprinted by permission of John Wiley & Sons, Inc.

Figure 4.7, page 107: From *Carbon-13 NMR Spectroscopy*, third edition, by E. Breitmaier and W. Voelter, Figure 3.3 ("$^{13}$C Chemical Shift Ranges in Organic Compounds"). Copyright © 1987 by VCH Publishers, New York, New York.

Figure 5.2, page 122: From *Interpretation of Mass Spectra*, third edition, by F. W. McLafferty, Figure 1.C ("Single Focusing, Magnetic-Sector Mass Spectrometer"). Copyright © 1980 by University Science Books, Mill Valley, California.

Appendix A, page 153: Adapted from *Spectrometric Identification of Organic Compounds*, third edition, by R. N. Silverstein, G. C. Bassler, and T. C. Morrill; Appendix A of Chapter 2 ("Masses and Isotopic Abundance Ratios for Various Combinations of Carbon, Hydrogen, Nitrogen, and Oxygen"), pages 41–67. Copyright © 1974 by John Wiley & Sons, Inc. Reprinted by permission of John Wiley & Sons, Inc.

# Note to the Instructor

*Interpreting Spectra of Organic Molecules* was written to be used in several different ways. For an introductory or short course on spectroscopic methods in organic chemistry, the text may be supplemented with problems from other sources, including spectra that students have recorded in a laboratory course. Second, this book may be used in conjunction with standard organic chemistry textbooks, most of which contain spectroscopy problems within each chapter. For both of these applications, this book provides the important concepts and strategies needed for a basic understanding of spectroscopic techniques in modern organic chemistry.

For a self-study module, this text is to be used in conjuction with the computer software *Spectroscopic Identification of Organic Compounds*. The software teaches students how to interpret spectra by presenting questions (and answers) that show what features of each type of spectrum are important and what information can be deduced. This book supplements the data presented in the software and also covers the underlying strategies that are necessary to interpret spectra. At the University of North Carolina, we use the book and software in conjuction with our second semester (sophomore) organic chemistry laboratory. At the beginning of the semester, teaching assistants cover the fundamentals of this text in two, one-hour lectures. The students then work with the software over the next several weeks (0 to 20 hours, depending on their interest) to reinforce their knowledge of the concepts and to learn how to interpret spectra. Toward the end of the course, each student identifies two unknown compounds based on their IR, NMR, and mass spectra.

# CHAPTER 1

# Strategies for Interpreting Spectra of Organic Molecules

Radiation of different wavelengths can cause changes in the electronic or molecular structures of organic compounds. The study that deals with the theory and interpretation of such interactions between molecules and radiant energy is called *spectroscopy*.

Figure 1.1 shows some of the different characteristics of electromagnetic radiation. Equations 1.1 to 1.3 summarize the relationships between energy ($E$), frequency ($v$), and wavelength ($\lambda$).

$$E = hv \qquad (h = 6.63 \times 10^{-27} \text{ erg sec}) \qquad (1.1)$$

$$v = c/\lambda \qquad (c = 2.998 \times 10^{10} \text{ cm sec}^{-1}) \qquad (1.2)$$

$$E = hc/\lambda \qquad (1.3)$$

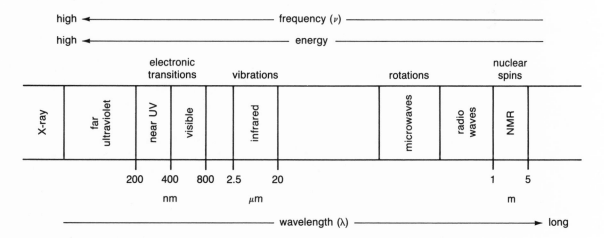

**Figure 1.1**
The electromagnetic spectrum. The three scales show the relative energies, frequencies, and wavelengths of radiation.

A spectroscopic analysis of a molecule requires at least one *spectrum*, which is a record of the changes that occur as you scan over the desired energy range. The most familiar type of spectrum results from plotting the change in the absorption of energy versus the wavelength of the energy being used. Other formats are sometimes more useful and will be described later. Rarely are you able to identify an organic compound from one spectrum. In practice, you will need several spectra, covering different energy ranges, to define a structure unambiguously.

A detailed presentation about each type of spectrum will be more informative if we first consider a general strategy for identifying organic molecules from spectroscopic data. Then each method can be viewed in relationship to the others, and you will be less tempted to overinterpret each spectrum. This chapter serves as the cornerstone for the book; please read it several times during the course of study to see how the overall strategy for spectral interpretation complements the information obtained from the individual spectra.

## I. ORGANIC MOLECULES

Before we look at the different types of spectra to see what kind of information is available from each, we will consider briefly the substances that we wish to identify, namely, the organic compounds themselves. Although you are undoubtedly familiar with these from your study of organic chemistry, you may never have stopped to consider what individual structural features actually constitute an organic molecule. Figure 1.2 presents structures of several diverse yet representative organic compounds for which four common features are notable:

(1) They all contain carbon, and there are usually bonds between two or more carbon atoms.

(2) Most contain hydrogen atoms, usually, but not always, bonded to carbon atoms.

(3) Many contain other elements, called heteroatoms, the most common being oxygen, nitrogen, sulfur, phosphorus, and the halogens. The heteroatoms can form bonds with each other or, more commonly, with carbon; and the bonds may be either single or multiple.

(4) Most contain a functional group, which is defined as any group of bonded atoms other than the carbon-carbon or carbon-hydrogen single bond. The functional groups range in complexity from the carbon-carbon double bond to multiply bonded heteroatoms.

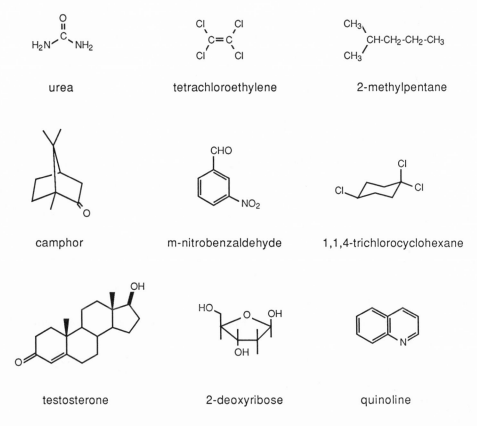

**Figure 1.2**
Representative structures of organic molecules.

## II. TYPES OF SPECTROSCOPY

Because organic compounds share the same few structural features, only a few spectroscopic techniques are needed to differentiate their structures. This book will be concerned with only the following four types of spectroscopy.

**a.** *Infrared (IR) spectra* result from absorption of energy that affects the vibrational modes of atoms that are bonded to one another. For a "typical" organic molecule, there should be many absorption bands, each produced by one of the many different groups of atoms. Since nearly all organic compounds contain C—C and

C—H bonds, many of the bands in an IR spectrum will be common to most substances and are therefore uninformative. As will be discussed later, IR spectroscopy is most useful for finding out what *functional groups* are present in a molecule.

   **b.** *Proton nuclear magnetic resonance (¹H NMR) spectra* result from absorption of energy that affects the spins of the hydrogen nuclei. The near-ubiquity of hydrogen atoms in organic structures makes NMR spectroscopy especially useful. You can use the information about the local environment of each set of hydrogen atoms to find out which carbon atom each hydrogen atom is attached to, and what functional group or heteroatom is nearby.

   **c.** *Carbon nuclear magnetic resonance (¹³C NMR) spectra* are analogous to proton NMR spectra, except that you can observe the carbon atoms directly. Since every organic compound contains carbon, you can use the ¹³C NMR spectrum to study the backbone of a molecule.

   **d.** *Mass spectra* result when a molecule is subjected to high-energy electron bombardment under vacuum and in a magnetic field. The molecule undergoes ionization, forming cation-radical species, followed by fragmentation of the molecular framework. The magnetic field is used to separate the different species based on their masses and charge (usually +1). As a result, mass spectroscopy is most useful for measuring the molecular weight of the compound and for identifying the presence of heteroatoms by their isotope patterns. In addition, certain carbon frameworks, for example, aromatic compounds, show characteristic fragmentation patterns.

   In summary, these four kinds of spectroscopy give you the following information about a compound:

   (1) the number and types of atoms present;

   (2) the environments of the carbon and hydrogen atoms;

   (3) the functional groups present.

This information parallels the essential structural features for organic molecules outlined in Section I.

   In learning more about each type of spectroscopy, try to develop a clear understanding of the advantages, disadvantages, and limitations of each method. Recognize, too, that you need not always use all four spectra to identify the structure of a molecule. In fact, you can often use either the IR and NMR or the NMR and ¹³C NMR spectra to deduce a compound's identity.

## III. GENERAL STRATEGIES FOR INTERPRETATION OF SPECTRA

The following chapters will cover the strategies for interpreting each type of spectrum in detail. The purpose of this section is to outline ways to use the kinds of spectroscopy listed in the preceding section to arrive at the structure of the molecule under study.

The first step in deducing the structure of an organic molecule is to consider the origin of the compound in question. Three typical situations are commonly encountered, and the strategy for applying the different spectroscopic methods to each one is presented in the following descriptions.

### A. Identification of a Predicted Reaction Product

The simplest situation in which spectroscopic identification of an organic compound is necessary is when the chemist makes a predictable change to a well-characterized compound. In this situation, you will not necessarily record all four types of spectra, because you can *predict* the changes that you should observe.

Since in all but the rarest cases, chemical transformations will change at least one functional group, you might expect to use IR spectroscopy most often for making sure that a desired reaction has occurred. However, although IR spectroscopy, or any other spectroscopic method, might be useful to indicate the progress of a reaction, you must still record other spectra to confirm that no other changes have taken place. Thus, you will almost always have to record the NMR spectrum to confirm that the carbon-hydrogen framework has remained intact or has undergone the appropriate changes, something for which IR spectroscopy is less well-suited. Because changes in functional groups will often affect the hydrogen-atom environments in predictable ways, NMR spectroscopy has actually become the technique used most often to monitor organic reactions.

Consider, for example, the reaction shown in Equation 1.4, in which a secondary alcohol is oxidized to a ketone.

$$\text{(1.4)}$$

You could use *any* of the four types of spectra to confirm that the desired reaction occurred: the molecular weight has decreased; the functional group has been altered substantially; and the number and environments of the protons are different. But which method will give the answer with the least work and ambiguity? Since the functional group has been altered, the IR and $^{13}C$ NMR spectra should show the most easily recognized changes (the changes that are actually observed will be discussed later). $^{13}$C NMR spectroscopy would also allow you to be sure that the other carbon atoms within the molecule have *not* undergone a reaction.

A second example illustrates what happens when a predictable reaction gives a product *mixture* as shown in Equation 1.5.

$$\text{(1.5)}$$

The first step in identifying the products here would be to isolate and purify the components. However, even when the two compounds have been isolated, spectroscopic identification is still necessary, not only to show that the desired reaction (i.e., functional-group change) has occurred, but also to distinguish between the two isomers. Again, as in the first example, any of the four spectroscopic methods will show that the reaction has occurred. The functional group has been changed from that of an alkyl bromide to that of an alkene (IR or $^{13}$C NMR spectra), the molecular weight has decreased by 80 mass units (mass spectra), and the number and type of protons have changed (NMR spectra). However, to distinguish *between* the two products, NMR spectroscopy would be the most useful, because the environments of the hydrogen atoms are quite different. For one isomer, the alkene carbon atoms are bonded to a methyl and a propyl group; in the other, the alkene carbon atoms are each bonded to an ethyl group. In contrast, the functional groups (*trans*-disubstituted double bond) and the molecular weights for the two isomers are the same; so the IR and mass spectra should be very similar.

## B. Identification of Reaction By-products

The second situation in which spectroscopic methods play an important role in organic chemistry involves the identification of an unexpected reaction product. Consider the reaction shown in Equation 1.6:

$$\text{(1.6)}$$

Here, you hope to carry out the substitution of the OH group in alcohol A to give bromide B, but find the additional product C. There are two options. The first is to treat product C as a complete unknown, and to follow the strategy outlined in the next section. The second option, which is often more desirable, is to use other information to narrow the possibilities. This other information would include:

(1) knowledge about the structure and reactivity of the starting material;

(2) the reaction conditions;

(3) the physical properties of the product.

For example, suppose you find that compound C is very nonpolar, and subsequently, from combustion analysis, that C contains only carbon and hydrogen. Since you know that alcohols tend to undergo an elimination reaction under acidic conditions, you might propose the following structure for C:

(1.7)

C

Now you can treat the problem as described previously, in which you confirm the structure of a known compound. Thus, the IR spectrum should indicate that the functional group has changed (OH vs. C=C); the mass spectrum should show that the molecular weight of the product is lower (122 vs. 104); and the NMR spectrum should show that all of the protons are bonded to $sp^2$ rather than $sp^3$ carbon atoms.

## C. Identification of an Unknown Compound

As shown by the last example, you can often use data on the physical properties and elemental composition of a compound to help identify its structure; but such information is not always available. For example, elemental analysis, which can confirm the presence of certain heteroatoms as well as the molecular formula, often requires large amounts of material in order to produce accurate results. If only a small amount of compound is available for identification, elemental analysis or wet chemical tests may waste valuable product. Using some combination of the four types of spectroscopy mentioned earlier will usually allow you to deduce the structure of an unknown compound without any additional information.

A strategy for identifying an organic molecule from its spectral data is outlined in Figure 1.3 (on page 8). The strategy is based on the premise that, at least for simple compounds, you can use proton NMR spectroscopy to establish the framework of a molecule, since most atoms in organic structures are either carbon or hydrogen. However, you need to know what heteroatoms and functional groups are present to make best use of the NMR spectrum.

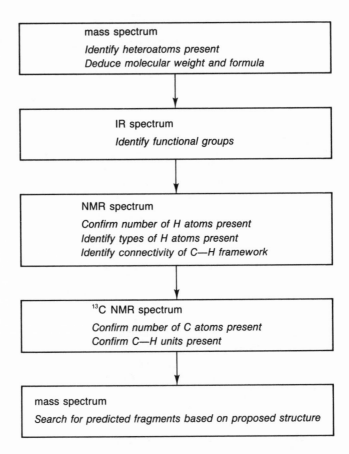

**Figure 1.3**
A flow chart summarizing the strategy for interpreting spectra of organic compounds.

You can begin by consulting the mass spectrum in order to find the molecular weight and the molecular formula. Since mass spectrometers are not always routinely available, the first step may be by-passed, or other methods may be used to find the molecular weight. Nevertheless, if you know the molecular formula, you can calculate the number of sites of unsaturation (pi-bonds or rings) using the formula:

$$\text{sites of unsaturation} = \frac{(2n + 2) - m}{2},\qquad (1.8)$$

where $n$ = number of carbon atoms in the molecule, and $m$ = number of hydrogen atoms in the molecule (add 1 to $m$ for each halogen atom; subtract 1 from $m$ for each nitrogen or phosphorous atom).

Subsequently you may be able to eliminate certain functional groups from further consideration. The mass spectrum will also indicate if any atoms of N, S, Cl, P, Br, or I are present.

Next, examine the IR spectrum to find out what major functional group(s) are present. You can use the $^{13}$C NMR spectrum to confirm the presence of functional groups that contain carbon atoms. Molecules having such groups include carbonyls (C=O), nitriles (C≡N), alkenes (C=C), and alkynes (C≡C).

Once you have identified the functional groups, examine the proton NMR spectrum and identify the different C—H fragments. The C—H groups can be fitted together, along with the known functionalities and heteroatoms, to generate the carbon skeleton with the attached hydrogen atoms and heteroatoms. Again, use the $^{13}$C NMR spectrum to ascertain that the proposed C—H fragments are in fact present.

Finally, you use the mass spectrum again to verify the proposed structure. Since you can often deduce the number of carbon atoms from the $^{13}$C NMR spectrum, and the number of hydrogen atoms from the proton NMR spectrum, you should be able to confirm the molecular formula. You can then predict probable fragmentation patterns based on the postulated structure, and see if the appropriate fragments appear with the correct mass-unit values.

The strategy we have outlined is not without pitfalls, although the identification of many molecules will yield to the approach summarized in Figure 1.3. There will obviously be problems if the compound contains no hydrogen atoms; one such compound is tetrachloroethylene. However, the mass spectrum would reveal the presence of the four chlorine atoms, and the $^{13}$C NMR spectrum would show a peak for the alkene carbon atoms. Another problem arises when there are many hydrogens in an aliphatic carbon framework, because, as you will see, they all appear in the same region of the NMR spectrum. For aliphatic compounds, interpretation of the mass spectrum can take on added importance.

As you learn more about each spectroscopic method, you will see the useful redundancies among the different techniques and learn what shortcuts you can take. Since there is no "right" way to analyze spectra, there are as many different ways to approach the interpretation process as there are practicing chemists. Those who have a strong background in mass spectroscopy feel most comfortable looking at the mass spectrum first. Similarly, chemists who work extensively with NMR spectra often turn to that technique first. This book presents only a few strategies for deducing the structures of organic molecules. As you accumulate your own experiences interpreting spectra, you should modify the strategy outlined in Figure 1.3 in ways that make it more logical for you.

# CHAPTER 2

# Infrared Spectroscopy

Infrared radiation has the energy needed to affect the stretching and bending vibrations of molecules. Organic compounds, with many covalent bonds, would be expected to have complicated, informative infrared spectra, and so they do.

The "normal" range for an infrared spectrum spans the wavelengths from 2.5 to 20 $\mu$m (micrometers). For simplicity, the units are usually expressed as $\mu$ (microns) instead of $\mu$m. It is even more common to express the positions of the peaks in an IR spectrum in units of frequency rather than of wavelength. Using Equation 1.2 to calculate the frequency range for an IR spectrum, you can see that the numbers are rather unwieldy, giving a range of $1.2 \times 10^{14}$ to $1.5 \times 10^{13}$ sec$^{-1}$ (or Hz). Because these frequency values are inconvenient, the term *frequency*, when applied to an infrared spectrum, is actually defined as $1/\lambda$ rather than as $c/\lambda$, and the units are expressed as cm$^{-1}$, termed *wavenumbers*, $\bar{v}$. The definition of wavenumbers, and the relationships for converting between wavelength and wavenumbers, are summarized in Equations 2.1 to 2.3:

$$\bar{v} = 1/\lambda \qquad (2.1)$$

$$\bar{v} = (10{,}000/\mu) \qquad (2.2)$$

$$\mu = (10{,}000/\bar{v}) \qquad (2.3)$$

You can see that the range of an infrared spectrum in wavenumbers spans 4000 to 500 cm$^{-1}$. Exponential and decimal notation are unnecessary with wavenumber values; so they are easy to use to express peak positions.

## I. THEORY

As with other types of energy absorption by molecules, absorption of infrared radiation is quantized; so energy of only certain frequencies can interact with the bonded atoms. Furthermore, in order for a band corresponding to a specific vibration to appear in the infrared spectrum, there must be a change in the dipole that accompanies the vibrational mode. Thus, a homobinuclear molecule like $H_2$ would not display a hydrogen-hydrogen stretch, since the dipole moment does not change if the $H-H$ bond is elongated or compressed. For organic molecules, having a center of symmetry inactivates any vibration encompassing the center point. Thus, symmetrically *trans*-disubstituted double bonds will not have a peak corresponding to the $C=C$ stretching mode. Internal alkynes and tetrasubstituted alkenes are also molecules with a center (or pseudo-center) of symmetry.

If you consider each chemical bond of an organic molecule as two balls with masses $m_1$ and $m_2$ connected by a spring with a force constant of $k$, then you can treat each stretching vibration by the harmonic oscillator approximation given in Equation 2.4:

$$v = \frac{1}{2\pi} \frac{k}{\mu}, \tag{2.4}$$

where

$$\mu = \frac{m_1 m_2}{m_1 + m_2}$$

and $k$ = the force constant.

From this simple picture using balls and springs, we can draw two general, but useful, conclusions:

(1) The frequency of the vibration will be inversely related to the *masses* of the atoms bonded to one another. Thus, the heavier the atoms are, the lower the frequency of the vibration will be. The examples in (2.5) are illustrative:

$$
\begin{array}{cccccc}
C=O & \text{vs.} & C=S & & C-H & \text{vs.} & C-D \\
\bar{v} = 1700 \text{ cm}^{-1} & & 1350 \text{ cm}^{-1} & & 3000 \text{ cm}^{-1} & & 2200 \text{ cm}^{-1}
\end{array}
\tag{2.5}
$$

(2) The frequency of the vibration will be directly proportional to the *strength* of the bond (the force constant). As expected, then, stretching vibrations of triple bonds will appear at higher frequencies than those of either double or single bonds:

$$
\begin{array}{ccc}
C\equiv C & C=C & C-C \\
\bar{v} = 2150 \text{ cm}^{-1} & 1650 \text{ cm}^{-1} & 1200 \text{ cm}^{-1}
\end{array}
\tag{2.6}
$$

The preceding examples assumed that each set of atoms, along with its bond, acts independently of all others in the molecule. For an aldehyde group, the assumption is reasonable, because the bond-force constants and masses of the atoms are quite different; so the $C-H$, $C=O$, and $C-C$ stretching vibrations will be relatively independent. However, for a methyl group, the three hydrogen atoms all have the same mass, and the $C-H$ bonds are all of similar strength; so the vibrations will not be independent. Instead, the vibrational modes are coupled, and appear as the symmetric and asymmetric $-CH_3$ stretching vibrations:

$$\nu_{C\text{-}H\ sym} = 2872\ cm^{-1} \qquad\qquad \nu_{C\text{-}H\ asym} = 2962\ cm^{-1} \qquad (2.7)$$

A very common set of coupled vibrations occurs with a group of three atoms. Examples include the methylene, nitro, and amino groups:

$$\qquad\qquad\qquad\qquad\qquad\qquad\qquad\qquad\qquad\qquad\qquad\qquad (2.8)$$

These will have two bands, corresponding to the symmetric and the asymmetric stretching modes. Coupled vibrations for the three-atom system provide evidence for the presence of those substructures within a molecule.

Most of the vibrational modes that will be discussed in Section III are the *fundamental vibrations*. These occur when the molecule absorbs IR radiation of the energy needed to promote it from the ground state to the first vibrationally excited state. However, other bands can occur that correspond to excitation to the second, third, or even fourth excited states. Such bands are called *overtones*; and although they are much weaker, they can be important for characterizing certain compounds, most notably aromatic molecules.

So far only stretching vibrations have been mentioned. In fact, many vibrations result from *bending* among bonded groups of three or more atoms. Bending modes are usually lower in energy than stretching vibrations, and bending modes between the heavier atoms give rise to bands at even lower frequency. Without going into detail, we will note that bending modes are described by such terms as scissoring, wagging, rocking, and twisting. (If you are interested in learning more about bending vibrations, consult the references listed at the end of this chapter.)

## II. EXPERIMENTAL CONSIDERATIONS

Very simply, infrared spectrometers, the instruments used to record IR spectra, operate by measuring the differences in energy between two beams of infrared radiation (heat), when one of them has passed through the sample. At frequencies where *no* absorption occurs, the beam that passes through the sample remains unaffected; so here the transmittance relative to the reference beam (the one that does not pass through the sample) is defined as 100 percent. Any absorption of radiation by the sample at a specific frequency results in a lower transmittance, leading to an absorption band or *peak* (try not to be confused by the fact that a "peak" is actually a dip in the line).

Figure 2.1 shows what a typical infrared spectrum recorded by a spectrometer looks like; the scan usually occurs from left to right. Note that the wavelength ($\mu$) of the radiation increases from left to right, but the frequency ($cm^{-1}$) decreases. The percent of transmittance of infrared energy through the sample is plotted so that decreased transmittance (or increased absorption) of radiation is toward the bottom of the spectrum.

There are three features of the IR spectrum on which you will need to focus, since each can give information that will lead to correct identification of structural elements:

(1) the position of the band (in wavenumbers);

(2) the shape of the band (broad, sharp);

(3) the intensity of the band (weak, medium, strong).

The position of the band is probably the most important feature, and one strategy for interpretation of the IR spectrum will be organized accordingly (Section III).

The spectrum overall can be divided into two general regions. The portion of the spectrum above $1300\ cm^{-1}$ ($1300–4000\ cm^{-1}$) is called the *functional group region* and will be the more important for interpretation. The other side of $1300\ cm^{-1}$ is called the *fingerprint region*, since the pattern of bands here is unique for each compound. The fingerprint region can be used for comparison, to match a spectrum of an unknown compound with spectra of known ones. A perfect match provides an unequivocal identification of a substance, just as matching fingerprints identify a person.

Before learning the strategies used to interpret infrared spectra, you need to know how a sample is prepared, since the state of the sample can affect the features of the IR spectrum. A pure compound is needed, so that all the bands observed can be attributed to it. However, the purity cannot always be ensured, since a medium must sometimes be added to allow the infrared beam to pass through the sample. Also, the sample will often absorb water from the air, giving bands around $3500\ cm^{-1}$. The sample container ("cell") should be transparent to infrared radiation. Materials like glass or plastic will

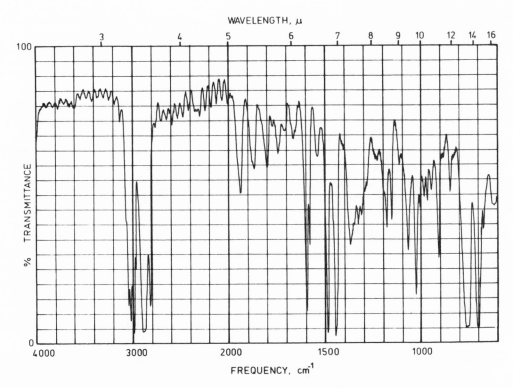

**Figure 2.1**
Infrared spectrum of polystyrene (thin film) showing the relationship between wavelength
and frequency.

have bands from Si—O or C—C and C—H bonds, respectively; so neither is very
useful. Usually cells are made of salts, which have no covalent bonds, with sodium
chloride, potassium bromide, and cesium fluoride being most often employed.

The type of sample determines what technique should be employed to obtain the
spectrum; so we will consider each type of sample in turn.

### a. *Gases*

The spectra of gases or very low-boiling liquids ($<25°$) are obtained by using a cell
made of a glass tube with salt plates at either end through which the IR beam can pass.
Such a cell is usually 10 cm long, since gaseous samples are relatively dilute. As with
absorption of UV or visible light, the absorption of IR radiation by a molecule often
follows Beer's Law ($A = \varepsilon bc$). Since the concentration of sample is low for gases at
atmospheric pressure, the path length of the cell must be long.

### b. *Liquids*

Several different methods can be used to record the spectrum of a liquid. If the compound boils at a fairly low temperature (20–60°), you can use a cell consisting of two salt plates surrounding a cavity in which the sample is placed. The path length for such cells ranges from 0.01 to 0.05 mm. Recall that IR radiation is just another name for heat; so volatile samples may evaporate during the recording of the spectrum.

For nonvolatile liquids, the spectrum is best recorded from a *thin film*. For this type of sample, a drop or two of the liquid is placed on a salt plate, and a second salt plate is pressed directly onto the sample, causing it to spread over the area between the plates, forming a thin film. This method is by far the easiest to carry out, and uses no other substances that might interfere with the appearance of the spectrum.

A third method for recording the spectra of liquids is to dissolve the compound in a solvent. That technique is covered in the following discussion of solid samples.

### c. *Solids*

Solids present the greatest problems in IR spectroscopy, because the infrared beam often cannot penetrate the sample far enough to give well-resolved spectra. Hence other media must often be used, each of which can interfere with the spectrum of the compound under investigation.

Low-melting solids can sometimes be examined if you place a small amount of compound between two salt plates, and rely on the heat from the IR beam to keep the sample molten, forming a thin film as described for liquids.

More often, a *mulling agent* must be used. There are two types. The first type is oily substances, such as mineral oil (often referred to as a "Nujol mull"), or halogenated hydrocarbons, such as hexachlorobutadiene or Fluorolube. Prepare the sample by grinding several milligrams of it in a mortar made of agate; add one or two drops of the mulling agent, and continue grinding. The oily suspension is spread between two salt plates, and the spectrum recorded as for a thin film. You must take the bands resulting from the oil (Figure 2.2) into account when you examine the spectrum, and some regions may be obscured.

A second type of mulling agent is a salt, usually potassium bromide, KBr. In this technique, 1 to 2 mg of the compound and 50 to 100 mg of KBr are ground together to make an intimately mixed fine powder. The powder is pressed at 20,000 psi under vacuum to form a transparent disk or *pellet*, which is then placed in the infrared beam in the spectrometer. Because KBr lacks any vibrational modes in the normal range used to study organic compounds, no extra bands are observed. Why is any other technique ever used to record IR spectra of solid samples? Unfortunately, KBr is somewhat hygroscopic, and O—H stretching modes from water are often observed around 3500 cm$^{-1}$. Furthermore, if the compound being studied forms strong hydrogen bonds (alcohols, amines, acids), then the problem with water is exacerbated. Also,

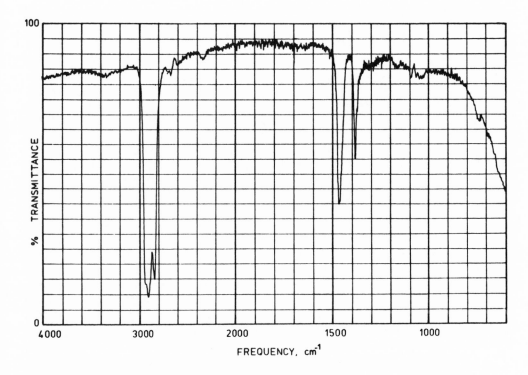

**Figure 2.2**
Infrared spectrum of mineral oil (thin film).

the KBr pellet is sometimes so opaque that the IR beam is not strong enough to give a reasonable spectrum.

A final technique for recording IR spectra of solids (and some liquids) is to use solvents that will give a clear solution. The compound is dissolved to give a concentration of usually less than 10 percent. A cell consisting of two salt plates separated by a spacer is filled with the solution, and the spectrum is recorded. Again, as with liquid mulls, the bands resulting from absorption by the solvent must be taken into account.

Carbon tetrachloride, although mediocre at dissolving many organic molecules (and a known carcinogen), is a good solvent to use for IR spectra, because the C—Cl stretches and bends occur at low frequencies, and there are no bands in its spectrum above about 1350 cm$^{-1}$ (1350–4000 cm$^{-1}$). Conversely, carbon disulfide shows few absorptions *below* about 1350 cm$^{-1}$. Other solvents are not as transparent; so you might have to use several to examine the entire IR range without interference from the medium. (Obviously, you cannot use a solvent that reacts with the compound being studied.)

## III. INTERPRETATION OF THE INFRARED SPECTRUM

Since IR spectra often contain more than fifty absorption bands, many of which are combinations of different stretching and bending modes, you should be aware that many of the peaks in the spectrum will not be considered in the interpretation process. Although we are here presenting a systematic approach to interpreting an infrared spectrum, there is not just one correct way to go about examining the IR spectrum.

There are two strategies presented below: the "Frequency Assignment Approach" (Section III.A), and the "Heteroatom Approach" (Section III.B). Each will be discussed separately, although in practice you might use a combination of the two.

### A. The Frequency Assignment Approach

Recall from Chapter 1 that from the interpretation of an infrared spectrum, you acquire information about the functional group(s) within a molecule. Since much of the molecule is the carbon and hydrogen skeleton, many of the bands in a spectrum will not help you identify the functional groups. Therefore you first examine bands appearing in regions *that are blank in the spectra of simple hydrocarbons*, since those bands can be correlated with the presence of the functional groups. After the functional groups have been identified, then you can examine bands corresponding to the C—H framework.

The *frequency assignment approach* is the most general strategy you can use, because it requires no other information about the molecular structure in order to make accurate, preliminary assignments. The overall strategy is shown in Figure 2.3.

#### 1. The Carbonyl Stretch Region (1800–1650 cm$^{-1}$)

The first place to look in any infrared spectrum is the region around 1700 cm$^{-1}$, where the carbonyl (C=O) stretch occurs. Since the C=O stretching mode is *always* intense, because of the large dipole moment of the carbon-oxygen double bond, you will rarely confuse it with any other absorption. Moreover, if no band is observed in the region around 1700 cm$^{-1}$, then immediately you have eliminated several classes of compounds from further consideration (see Figure 2.3).

Table 2.1 (on page 20) lists the possible carbonyl-containing functional groups, along with the stretching frequencies associated with aliphatic and aromatic (or $\alpha,\beta$-unsaturated) compounds containing the C=O group. Figure 2.4 (on page 21) shows the same data graphically.

The differences in the C=O stretching frequency of the various carbonyl groups can be ascribed to a combination of resonance and inductive effects. (We will use the C=O stretch of an aliphatic ketone as the reference point in the following discussion, since its carbonyl group is attached to two saturated carbons.) When a strongly

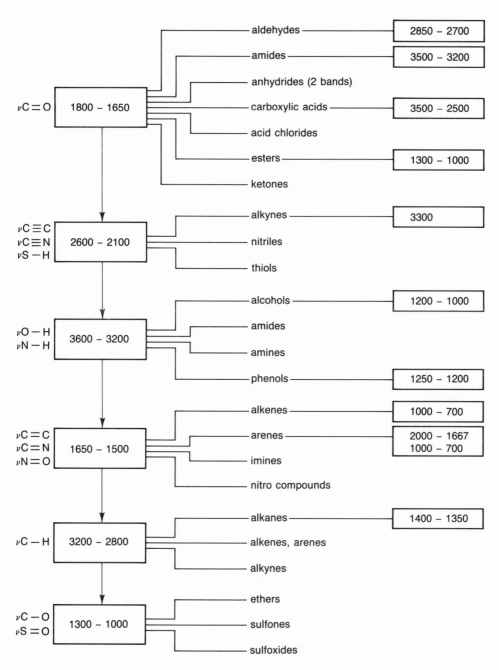

**Figure 2.3**
A systematic strategy for interpreting an infrared spectrum. All values are in cm⁻¹.

**Table 2.1**
Frequency ranges for the $C{=}O$ stretching vibration in carbonyl compounds[a]

| Compound type | Saturated | $\alpha,\beta$-Unsaturated and aryl |
|---|---|---|
| Acid chloride (RCOCl) | 1815–1790 | 1790–1750 |
| Aldehyde (RCHO) | 1730–1710 | 1715–1680 |
| Anhydride (RCO—O—COR) | 1850–1800<br>1790–1740 | 1830–1780<br>1770–1710 |
| Amide, 1° (RCONH$_2$) | $\sim 1650$<br>$\sim 1640$ }[b]<br>1700–1685<br>1605–1595 }[c] | |
| Amide, 2° (RCONHR′) | 1680–1630<br>1570–1515 }[b]<br>1700–1670<br>1550–1510 }[c] | |
| Amide, 3° (RCONR$_2'$) | 1670–1630 | |
| Carboxylic acid (RCOOH) | 1725–1700 | 1715–1680 |
| Ester (RCOOR′) | 1750–1735 | 1730–1715 |
| Ketone (RCOR′) | 1725–1705 | 1700–1660 |

[a] All values given in cm$^{-1}$.
[b] Solid-state spectrum.
[c] Solution spectrum.

electronegative atom is attached to the carbonyl carbon, the inductive effect is dominant, and strengthens the $C{=}O$ bond because of partial donation of the carbonyl-oxygen lone pairs to the carbonyl-carbon atom. Thus acid chlorides, anhydrides, and, to a lesser extent, esters have their carbonyl absorptions at *higher* frequencies than the carbonyl absorption of ketones:

(2.9)

On the other hand, amides, even though the carbonyl group is attached to an electronegative atom, have $C{=}O$ stretching frequencies *lower* than the ketone value, because resonance stabilization from the nitrogen atom decreases the amount of double-bond character in the carbonyl group, thereby weakening the bond:

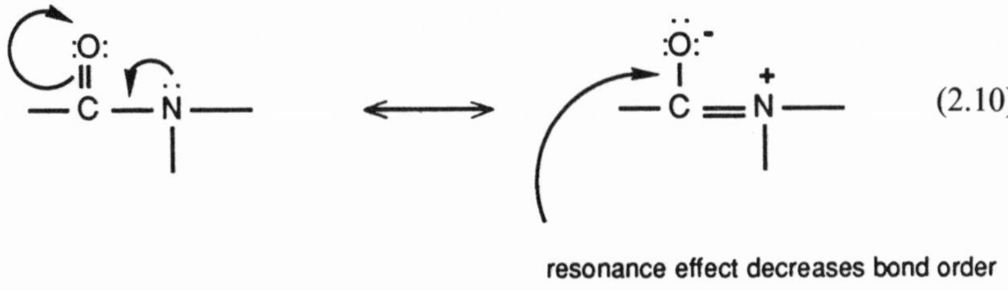

(2.10)

resonance effect decreases bond order

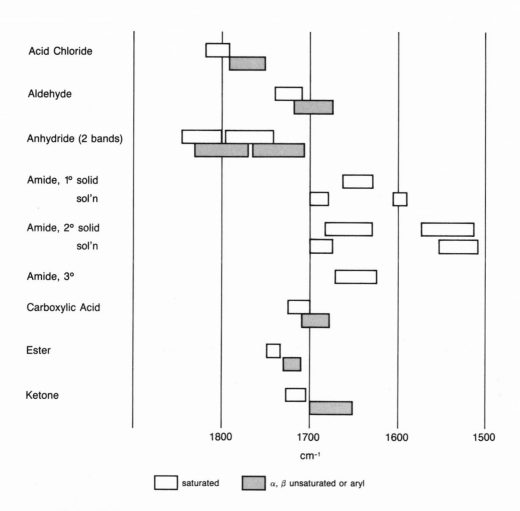

**Figure 2.4**
Frequency ranges of bands in the C═O stretch region for carbonyl compounds.
All bands are strong.

Note that two bands are actually observed for 1° and 2° amides. The bands are termed the "amide I band" ($C=O$ stretch) and the "amide II band" (largely $N-H$ bending and deformation).

Conjugation of the carbonyl group with the $C=C$ bond of alkenes or arenes also influences the $C=O$ group, by diminishing its double-bond character. Thus conjugated carbonyl compounds have their $C=O$ absorptions at lower frequencies than the $C=O$ stretch observed for their saturated counterparts (see Table 2.1):

$$\begin{array}{ccc}
\overset{\displaystyle O}{\underset{\displaystyle \parallel}{}} & & \overset{\displaystyle O^-}{\underset{\displaystyle |}{}} \\
-C-C=C & \longleftrightarrow & -C=C-\overset{+}{C} 
\end{array} \qquad (2.11)$$

An exception to the last statement is provided by compounds having the amide group. Because the delocalization of the carbonyl $\pi$-bond with the amide nitrogen atom minimizes delocalization with the $C=C$ $\pi$-system, the $C=O$ stretch for unsaturated amides is not very different from the $C=O$ absorption observed for saturated amides:

$$\begin{array}{ccc}
& & \overset{O^-}{|} \\
& & -\overset{+}{N}=C-C=C \\
\overset{O}{\parallel} & \nearrow & \text{major contributor} \\
-N-C-C=C & & \\
& \searrow & \overset{O^-}{|} \\
& & -N-C=C-\overset{}{\underset{+}{C}} \\
& & \text{minor contributor}
\end{array} \qquad (2.12)$$

Besides the variation in stretching frequency resulting from the inherent differences among the functional groups, you also have to be alert for variations that result from steric effects, especially changes generated by ring strain (Figure 2.5) and hydrogen bonding. Ring strain usually compresses the $C-CO-C$ angle at the carbonyl carbon; in a strain-free environment, it is 120°. Ring strain results in increased s-character at the carbonyl carbon, which in turn strengthens the $C=O$ bond because of increased overlap between orbitals on the carbon and oxygen atoms. Consequently, the force constant of the bond is greater, and the frequency of the absorption *increases*.

Hydrogen bonding causes a *lowering* of the carbonyl stretching frequency, because mutual interaction of the oxygen and hydrogen atoms results in lengthening of the $C=O$ bond, and the force constant of the $C=O$ bond is decreased. Because of hydrogen bonding, the carbonyl group of carboxylic acids (which exist primarily as

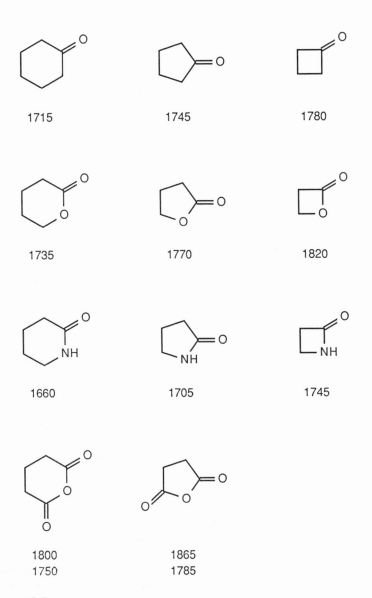

**Figure 2.5**
Effect of ring size on the C=O stretching frequency for carbonyl compounds.
All values are in cm$^{-1}$.

dimers except in very dilute solution) absorbs at lower frequencies than the carbonyl group of structurally analogous esters:

$$\nu_{CO} = 1710 \ cm^{-1}$$

$$\nu_{CO} = 1735 \ cm^{-1}$$

(2.13)

$$\nu_{CO} = 1720 \ cm^{-1}$$

$$\nu_{CO} = 1680 \ cm^{-1}$$

The absorptions for 1° and 2° amides are similarly affected by hydrogen bonding, as well as by whether the spectrum has been recorded on a sample in the solid or solution phase (see Figure 2.4).

The preceding discussion shows that there is much overlap between regions associated with the different carbonyl-containing functional groups. For example, a band at 1710 cm$^{-1}$ could result from an aliphatic ketone, a carboxylic acid, an aromatic ester, or an α,β-unsaturated aldehyde. Clearly, you cannot arrive at a dependable conclusion given only the carbonyl stretching frequency. If you are interested in finding out exactly which type of molecule is associated with the carbonyl band, you will need more data in order to distinguish the carbonyl-containing groups from one another. That information is summarized in Table 2.2. Note that sometimes the IR spectrum alone will not enable you to distinguish between two functional groups; therefore you will need other spectra in order to characterize organic molecules completely. As additional data about the molecular structure becomes available, either from closer examination of the IR spectrum or from other spectra, check the assignments made early in the interpretation process to make sure that the conclusions are still valid.

**Table 2.2**
Distinguishing features for carbonyl compounds[a]

| Compound type | $\nu CO$[b] | Principal features to look for: |
|---|---|---|
| Acid chloride | 1800 | single C=O stretch near upper end of the carbonyl range; presence of chlorine (mass spectrum); see Figure 2.6. |
| Aldehyde | 1725 | C—H stretch near 2720 (medium intensity); proton resonance near 10 ppm in NMR spectrum; see Figure 2.7. |
| Anhydride | 1820 1760 | two bands in the carbonyl region; broad and strong C—O stretch 1300–1000; see Figure 2.8. |
| Amide (1°, 2°) | 1680 | N—H stretch around 3300; two bands in the carbonyl region (C=O stretch and N—H bend); presence of nitrogen (mass spectrum); see Figures 2.9 and 2.10. |
| Amide, 3° | 1650 | single C=O stretch near lower end of the carbonyl range; presence of nitrogen (mass spectrum); see Figure 2.11. |
| Carboxylic acid | 1710 | O—H stretch between 3500 and 2500 (very broad and strong); see Figure 2.12. |
| Ester | 1735 | strong, broad band around 1200, sometimes more intense than C=O stretch; see Figure 2.13. |
| Ketone | 1715 | no outstanding features; see Figure 2.14. |

[a] All values given in $cm^{-1}$.

[b] Base value for saturated compound; see Table 2.1 for ranges.

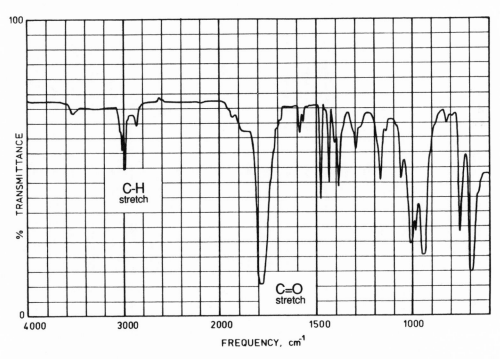

**Figure 2.6**
Infrared spectrum of phenylacetyl chloride (thin film).

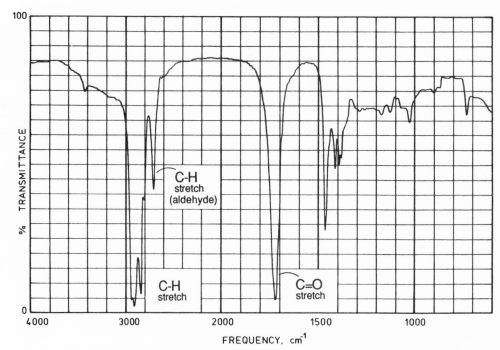

**Figure 2.7**
Infrared spectrum of nonanal (thin film).

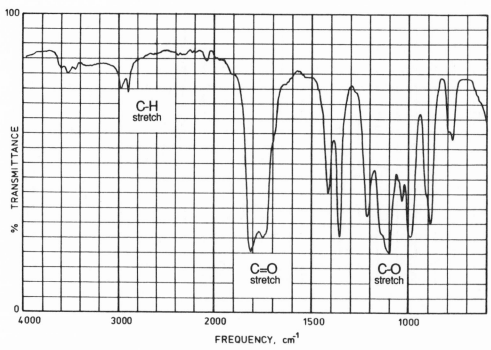

**Figure 2.8**
Infrared spectrum of acetic anhydride (thin film) showing the two C=O stretch bands.

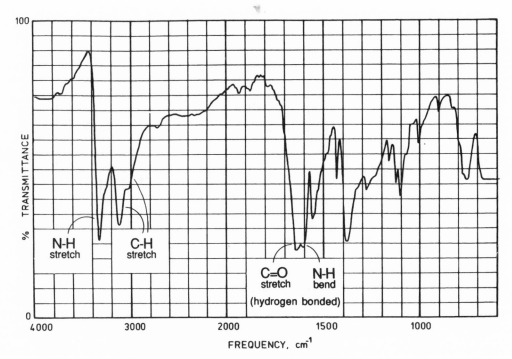

**Figure 2.9**
Infrared spectrum of benzamide (KBr pellet).

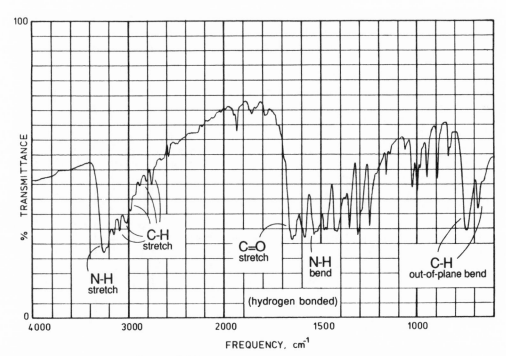

**Figure 2.10**
Infrared spectrum of acetanilide (KBr pellet).

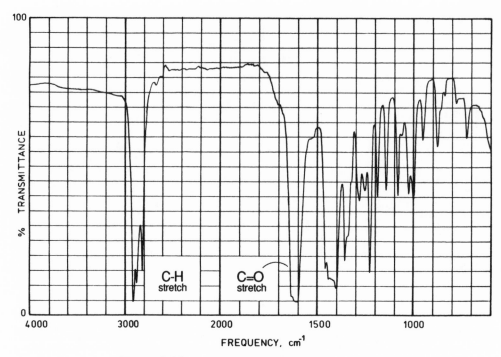

**Figure 2.11**

Infrared spectrum of *N*,*N*-dipropylacetamide (thin film).

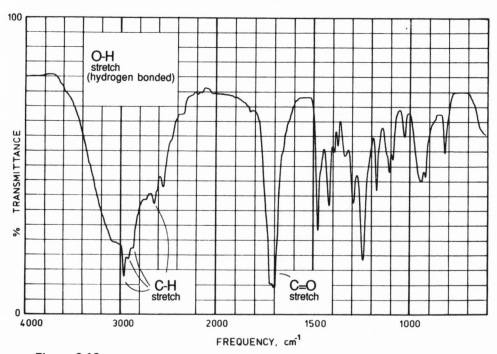

**Figure 2.12**

Infrared spectrum of isobutyric acid (thin film) illustrating the broad O — H stretch band resulting from intermolecular hydrogen bonding.

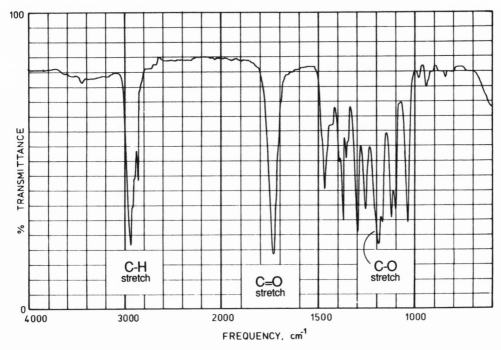

**Figure 2.13**
Infrared spectrum of ethyl 3-methylbutanoate (thin film).

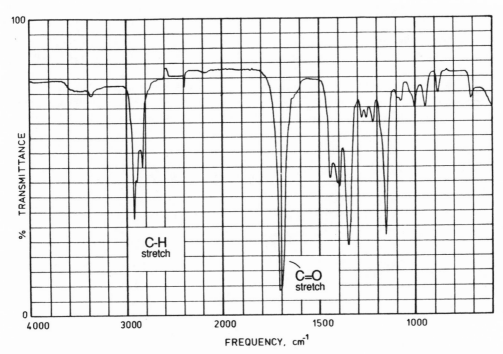

**Figure 2.14**
Infrared spectrum of 2-pentanone (thin film).

### 2. The Triple-Bond Stretch Region (2600–2100 cm$^{-1}$)

The next region to examine is the one in which triple bonds and S—H stretching vibrations appear (Table 2.3). The region between 2700 and 2000 cm$^{-1}$ is almost always blank, since relatively few molecules have those functional groups.

**Table 2.3**
Ranges of stretching frequencies for groups absorbing in the triple-bond region[a]

| Functional group | Saturated | $\alpha,\beta$-Unsaturated or aryl |
|---|---|---|
| Alkyne, terminal | 2150–2120 | 2140–2100 |
| Alkyne, internal | 2260–2190 | 2240–2150 |
| Nitrile | 2260–2240 | 2240–2220 |
| Thiol | 2580–2550 | 2600–2550 |

[a] All values in cm$^{-1}$; all intensities are variable.

Triple-bond stretching modes are of two types, C≡N and C≡C, and their absorptions usually occur between 2300 and 2100 cm$^{-1}$. The nitrile stretch is of variable intensity, but is usually observable because of the dipole associated with the CN group (Figure 2.15). Conversely, internal alkynes have very weak, often unobservable, bands, because the dipole for the C≡C bond is essentially zero (Figure 2.16). Recall that for a vibration to be observable, there must be a net dipole change when the bond interacts with infrared radiation.

In order to distinguish between alkynes and nitriles, you can first check the mass spectrum for the absence of nitrogen, which would eliminate the nitrile group from further consideration. If you think that the molecule contains a terminal alkyne group (≡C—H), then you can check for a sharp, medium-intensity peak at about 3300 cm$^{-1}$ that results from the alkyne C—H stretch (Figure 2.17, page 32). Finally, the $^{13}$C NMR spectrum may be of use here, since compounds with alkyne carbon atoms give spectra that are readily distinguishable from the spectra of compounds that have a nitrile carbon atom (Chapter 4).

The band that results from the S—H stretching vibration also appears in the region around 2500 cm$^{-1}$ (Figure 2.18, page 32); but since the S—H stretch appears at higher frequency, it is not easily confused with the triple-bond stretches. The mass spectrum can also be consulted to verify that the molecule contains sulfur.

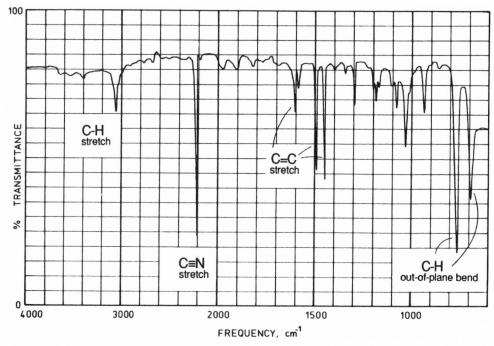

**Figure 2.15**

Infrared spectrum of benzonitrile (thin film).

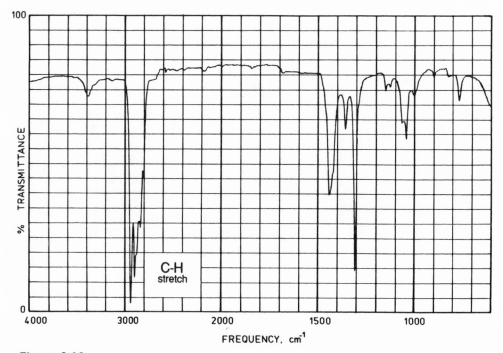

**Figure 2.16**

Infrared spectrum of 3-hexyne (thin film). Note that there is no C≡C stretch band in the region 2300–2100 cm$^{-1}$; *cf*. Figure 2.17.

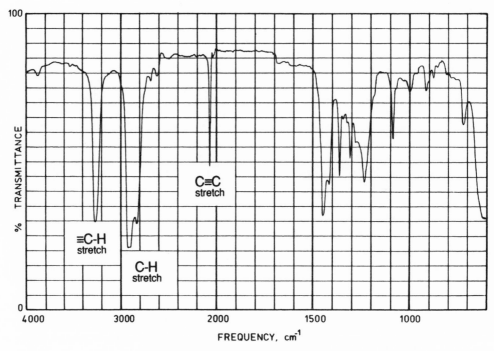

**Figure 2.17**
Infrared spectrum of 1-hexyne (thin film).

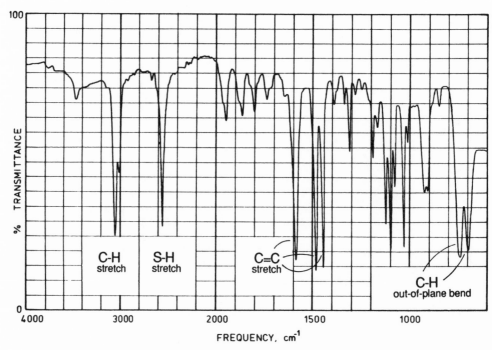

**Figure 2.18**
Infrared spectrum of benzenethiol (thin film).

### 3. The O — H and N — H Stretch Region (3600–3200 cm$^{-1}$)

After examining the triple-bond region, you can focus on the high-frequency region in which O — H and N — H stretching vibrations appear. You may see bands around 3500 cm$^{-1}$ that result from some water in the sample; so you need to make certain that the bands in this region belong to the compound. Because oxygen- and nitrogen-containing functional groups tend to undergo hydrogen bonding, bands associated with OH or NH groups in this region are often broad. You have already seen that the band for the O — H stretch of carboxylic acids begins around 3500 cm$^{-1}$ and extends to about 2500 cm$^{-1}$ (Figure 2.12). However, the band for the O — H stretch in alcohols and phenols is not as broad as that, even with hydrogen bonding. Table 2.4 summarizes the expected frequencies for bands in the region around 3500 cm$^{-1}$, along with other distinguishing features for each functional group.

**Table 2.4**
Summary of frequency ranges for OH and NH stretches by compound type[a]

| Compound type | OH or NH stretch[b] | Other diagnostic bands |
|---|---|---|
| Alcohol (H-bonded) | 3500–3200 (s, br) | C — O stretch: 1150–1000 |
| ("free" OH) | 3650–3600 (s, sh) | |
| Phenol (H-bonded) | 3500–3200 (s, br) | C — O stretch: 1250–1200 |
| ("free" OH) | 3650–3600 (s, sh) | |
| 1° amine | 3500–3300 (v, 2 bands) | N — H deformation: 1640–1560 |
| 2° amine | 3500–3300 (v) | N — H deformation: 1500 |
| 1° amide | 3500–3100 (v, 2 bands) | C═O stretch: 1700–1685 |
| 2° amide | 3500–3100 (v) | C═O stretch: 1700–1670 |

[a] All values are in cm$^{-1}$.
[b] s = strong, br = broad, sh = sharp, v = variable.

The first step is to decide whether the band in this region is a result of an O — H or N — H stretch. As a general rule, you can differentiate between alcohols and amines because, for concentrated samples, the N — H stretch is often sharper than the corresponding O — H mode (Figure 2.19, page 34). However, if you are able to record the spectrum with a very dilute sample, look for the free O — H stretch between 3600 and 3800 cm$^{-1}$ or the free N — H stretch between 3500 and 3300 cm$^{-1}$. The mass spectrum, if it is available, can also be helpful here, because it can indicate the presence or absence of nitrogen in the compound.

If the band in the region around 3500 cm$^{-1}$ is ascribed to the OH group, you can differentiate between the different types of alcohols and phenols by examining the C — O stretching region between 1250 and 1000 cm$^{-1}$. The expected values for the C — O stretching modes of the different hydroxyl-containing compounds are: phenols,

**34**

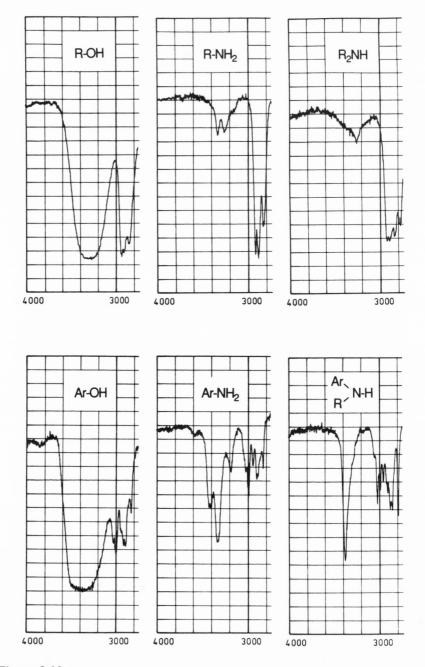

**Figure 2.19**
Comparison of OH and NH stretches for aliphatic and aromatic compounds: R—OH, 1-propanol; RNH$_2$, butylamine; R$_2$NH, dibutylamine; Ar—OH, *o*-cresol; ArNH$_2$, *o*-toluidine; Ar—(R)NH, N-methylaniline. The sharper bands around 3000 cm$^{-1}$ are the C—H stretching vibrations.

1220; 1° alcohols, 1150; 2° alcohols, 1100; and 3° alcohols, 1050; but those numbers are only "base values." Because the C—O stretch is coupled with neighboring C—C stretching modes, unsaturation of the carbon skeleton and steric effects, particularly ring strain, will affect the actual frequency.

On the other hand, if you think the band in the 3600–3200 cm$^{-1}$ region is an N—H stretch, you can usually differentiate between 1° and 2° amines, because primary amines will have two bands assigned to the symmetric and asymmetric stretches (recall from Section I that groups of three atoms like —NH$_2$ have two characteristic absorptions; see Figure 2.19). Secondary aliphatic amines often have a very weak N—H stretch, whereas the aromatic analogs have a more prominent peak (Figure 2.19). You can further clarify any ambiguity by examining the N—H bending region (1640–1560 cm$^{-1}$), because 2° aliphatic amines have no band in this region, whereas 1° amines will show a broad absorption. Since aromatic compounds also absorb here, the N—H bend may be obscured in the spectra of aromatic amines.

In addition to amines, an N—H stretch may result from the *amide* (or sulfonamide, RSO$_2$NH$_2$) functional group. Primary amides, like primary amines, show two strong absorptions (the symmetric and asymmetric stretches) above 3200 cm$^{-1}$; and those bands occur at about 3450 and 3225 cm$^{-1}$. Secondary amides will have only one band if the spectrum is recorded in dilute solution; however, the solid-state spectrum is often more complex (see Figure 2.10).

### 4. The Double-Bond Stretch Region (1650–1500 cm$^{-1}$)

The next region to examine is that in which C=C, C=N, and N=O stretches occur. The latter two are fairly rare among organic compounds; however, many molecules contain carbon-carbon double bonds. Table 2.5 summarizes the expected frequencies for functional groups appearing in the region around 1600 cm$^{-1}$.

**Table 2.5**

Frequency ranges for vibrations appearing in the double-bond region[a]

| Compound type | Bond | Frequency range[b] |
|---|---|---|
| Alkene | C=C | 1666–1640 (w-m) |
| Arene | C=C | 1650–1400 (v) (3–4 bands) |
| Imine | C=N | 1690–1640 (m) |
| Nitroso | N=O | 1600–1500 (s) |
| Nitro | N=O | 1600–1500 (s) 1390–1300 (s) |

[a] All values in cm$^{-1}$.
[b] w = weak, m = medium, s = strong, v = variable.

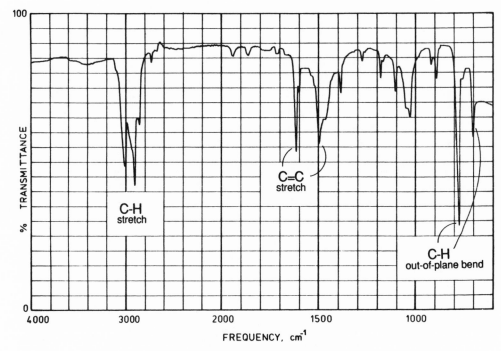

**Figure 2.20**
Infrared spectrum of *m*-xylene (thin film).

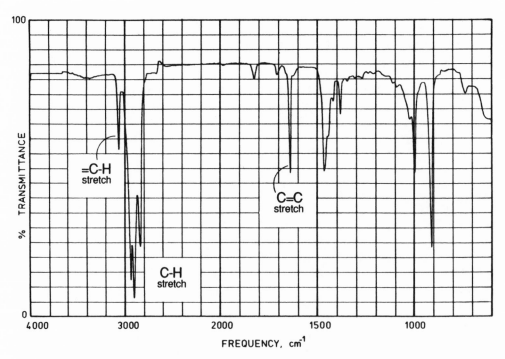

**Figure 2.21**
Infrared spectrum of 1-hexene (thin film).

The $C≡C$ bond gives rise to a variable-intensity band. For alkenes with a center of symmetry, the $C≡C$ stretch may be so weak that it is unobservable, since there is no net dipole change during the infrared absorption process. *Trans*-disubstituted and tetrasubstituted double bonds are compounds in which no dipole change occurs. Although aromatic compounds lack formal $C≡C$ bonds, they also absorb around 1600 cm$^{-1}$; and the intensities of the $C≡C$ stretching modes in arenes are variable, too.

Distinguishing between alkenes and arenes is usually straightforward, since several regions of their IR spectra show noticeable differences. Nonconjugated alkenes have their $C≡C$ stretch at higher frequencies than most aromatic compounds (Figures 2.20 and 2.21), but for conjugated alkenes, the $C≡C$ stretch occurs at lower frequencies. For the latter, the delocalized $\pi$-system results in more single-bond character at the alkene carbon atoms, leading to a weaker force constant for the $C≡C$ bond, and hence a lower frequency for the $C≡C$ stretching vibration.

$$C=C-C=C \quad\longleftrightarrow\quad \overset{-}{C}-C=C-\overset{+}{C} \qquad (2.14)$$

decreased double-bond character

Note, too, that ring strain can affect the frequency at which you observe the $C≡C$ stretch (Figure 2.22). Therefore, although the $C≡C$ stretch by itself may enable you to distinguish between an alkene and arene, you may find it worthwhile to check other regions of the spectrum as well.

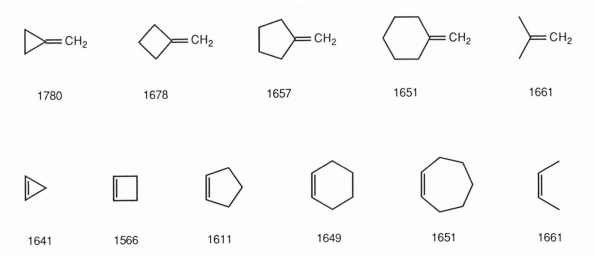

**Figure 2.22**
Effect of ring size on the $C≡C$ stretching frequency for both exocyclic and endocyclic double bonds. All values are in cm$^{-1}$.

38

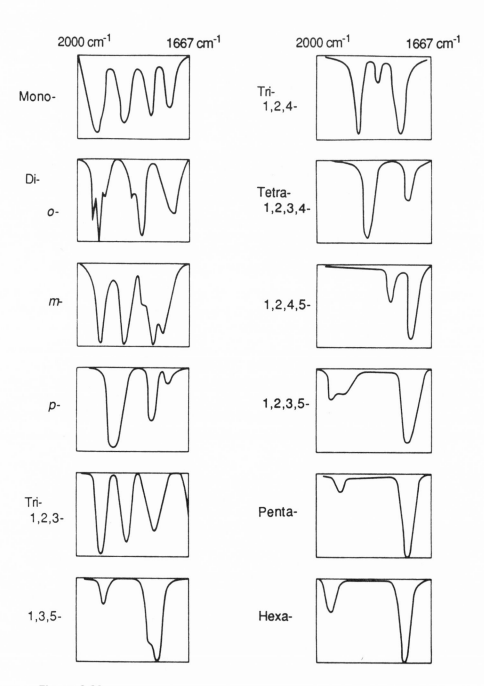

**Figure 2.23**
Absorption patterns in the overtone region for substituted benzenoid compounds.

The second region that you might use to help distinguish between alkenes and arenes is the so-called "overtone" region between 2000 and 1667 cm$^{-1}$. Because overtone bands are often weak, you will need to have concentrated samples in order to record useful spectra of this region. The overtone region can be used to ascertain the substitution pattern of a benzene derivative, as shown in Figure 2.23. Unfortunately, if the molecule contains a carbonyl group, the C=O stretch will often block enough of the overtone region to limit its usefulness in assigning ring-substitution patterns.

The third region that can be used to distinguish alkenes from arenes is the "C—H out-of-plane bend" region between 1000 and 650 cm$^{-1}$. Absorptions here result from the motion shown in (2.15), and the data are summarized in Figure 2.24 (on page 40).

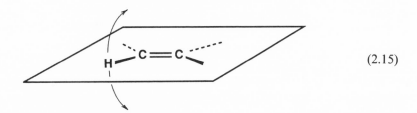

$$(2.15)$$

Although there is some overlap between the different types of molecules that absorb in the region between 1000 and 650 cm$^{-1}$, the spectra are often unambiguous, and can be used to strengthen a conclusion drawn from examination of the C=C stretch region. Not only is Figure 2.24 useful as a guide for deciding whether a compound is an alkene or arene, but you can deduce the substitution pattern of the C=C bond or the phenyl ring from the pattern of absorptions between 1000 and 600 cm$^{-1}$.

The final way to distinguish between an alkene and arene is to use the NMR spectrum, since often the protons will absorb in slightly different regions of the spectrum. The NMR spectra of arenes and alkenes will be discussed in Chapter 3.

Besides compounds having a C=C stretching vibration, two other types of compounds absorb in the region between 1650 and 1500 cm$^{-1}$, namely, compounds with C=N and N=O groups. The nitroso (—N=O) and nitro (—NO$_2$) moieties are fairly simple to detect, because their bands are very intense, like absorptions for C=O compounds. Furthermore, the nitro group, which is the more common, has two bands (Figure 2.25, page 41): a symmetric stretch between 1600 and 1500 cm$^{-1}$, and an asymmetric stretch between 1390 and 1300 cm$^{-1}$ (recall the three-atom arrangement discussed in Section I). The C=N group, because its absorption occurs in the same region as C=C bonds (Table 2.5), can be missed unless you have the mass spectrum to indicate the presence of nitrogen in the molecule. Fortunately, C=N bonds are rare in most organic molecules.

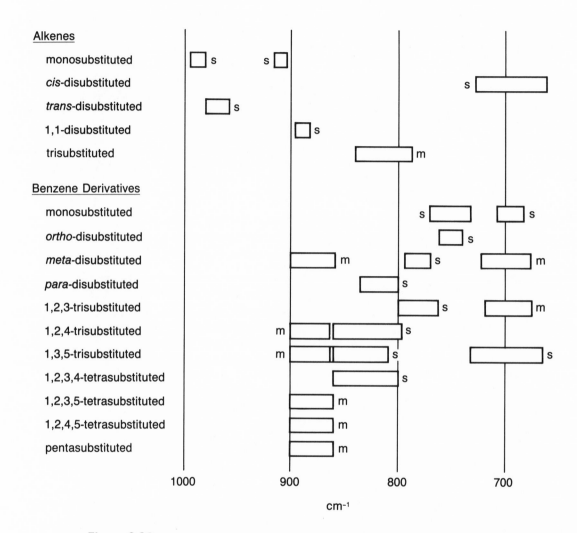

**Figure 2.24**
Frequency ranges for the C—H out-of-plane bending vibrations for substituted alkenes and benzenes; s = strong, m = medium intensity.

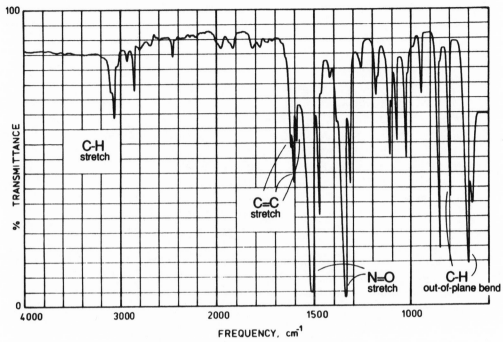

**Figure 2.25**
Infrared spectrum of nitrobenzene (thin film).

## 5. The C—H Stretch Region (3300–2700 cm⁻¹)

Because almost every organic molecule contains hydrogen, the region around $3000 \text{ cm}^{-1}$ is almost certain to have bands from C—H vibrations. Hydrocarbon absorptions, therefore, provide little useful information about the structure. Table 2.6 summarizes the different C—H bonds that a molecule may have and their associated stretching frequencies.

**Table 2.6**
Frequency ranges for C—H stretching vibrations

| C—H bond type | C—H stretch (cm⁻¹) |
|---|---|
| ≡C—H | 3330–3260 |
| =C—H | 3200–3000 |
| ▷—H | 3100–2990 |
| ⩵C—H | 3000–2840 |
| —C—H<br>‖<br>O | 2850 and 2750[a] |

[a] Sometimes not observed in aliphatic compounds.

The unusual C—H bonds will be addressed first, since you have already encountered them. The ≡C—H stretch occurs at the highest frequency, and the aldehyde C—H stretch at the lowest. For each of these functional groups, you would have already looked for the C—H vibrational band when examining the triple-bond or carbonyl region, respectively.

The bands that are left arise from vibrations of $(sp^2)$carbon- or $(sp^3)$carbon-hydrogen bonds. Immediately you should see that, in general, the former occur above $3000 \, cm^{-1}$ ($3200-3000 \, cm^{-1}$), and the latter (with the exception of cyclopropyl) below $3000 \, cm^{-1}$ ($3000-2800 \, cm^{-1}$). Thus you have a simple way to distinguish between aromatic and aliphatic compounds. Substitution of the aromatic ring can cause some arene C—H stretching bands to appear below $3000 \, cm^{-1}$.

Observation of absorption bands above $3000 \, cm^{-1}$ should merely provide confirmation of the alkene or arene assignment that you made during examination of the C=C stretch region ($1650-1500 \, cm^{-1}$). However, bands between 3000 and $2800 \, cm^{-1}$ provide the first indication during the interpretation process that the molecule has an aliphatic portion.

*Infrared spectra are not well-suited to probing the structure of a saturated hydrocarbon in detail, but are more useful for identifying functional groups.* However, you can sometimes obtain useful information about the hydrocarbon framework from the IR spectrum. If you decide from the C—H stretch region that the molecule has an aliphatic portion, then the next region to examine is around $1380 \, cm^{-1}$, where the methyl symmetric-bending mode occurs. The appearance of peaks near $1380 \, cm^{-1}$ can sometimes define the environment that the methyl groups occupy. The expected patterns are illustrated in Figure 2.26, but the observed pattern may differ substantially from the idealized ones presented here. Fortunately, any conclusions about the environments of the methyl groups can often be verified by using NMR spectroscopy, which focuses directly on the hydrogen-atom environments.

The final region to examine is near $700 \, cm^{-1}$, where the methylene "rocking" vibrational mode occurs. The intensity of this band is proportional to the number of methylene groups in a hydrocarbon chain; so a band of medium to high intensity indicates a long-chain compound. However, several C—H out-of-plane bends occur around $700 \, cm^{-1}$ as well; so a firm conclusion cannot be based on an absorption near $700 \, cm^{-1}$ if the molecule has an aromatic or alkenyl fragment in addition to the aliphatic group.

### 6. The Fingerprint Region ($1300-600 \, cm^{-1}$)

The last region to examine constitutes the rest of the spectrum. As was stated in Section II, the fingerprint region is most useful for matching the spectrum of an unknown compound with spectra of known substances. A perfect correspondence of peaks is definitive proof of structure, in the same way that matching fingerprints identify a person. Interpretation of bands in the fingerprint region is complicated,

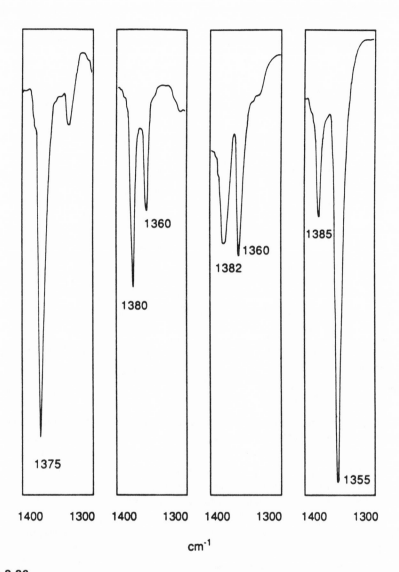

**Figure 2.26**

Symmetric methyl bending vibrations for (a) a single methyl group (octane); (b) a gem-dimethyl group (isobutyryl chloride); (c) a gem-dimethyl group (isobutanol); (d) *tert*-butyl group (2,2-dimethylpropanonitrile). The separation between the peaks is generally greater for the *t*-butyl group versus the gem-dimethyl moiety, and the lower frequency peak is generally more intense for the *t*-butyl group. Intensities in this region for the gem-dimethyl group can differ, as shown by the two examples here.

because many absorptions result from the carbon backbone of the molecule and are combinations of several vibrational modes. However, several strong peaks may be identified *as long as you proceed cautiously*. Notable absorptions are summarized in Figure 2.27.

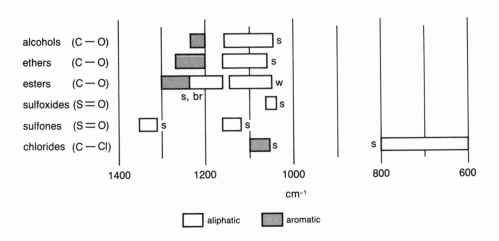

**Figure 2.27**
Frequency ranges for principal bands appearing in the fingerprint region; s = strong, w = weak, br = broad.

The ether functionality has C—O stretches that absorb in the region between 1300 and 1000 cm$^{-1}$. You must be certain that the C—O stretch is not the alcohol or ester C—O stretch; for each of the latter, an additional band will be observed, namely, the O—H stretch or the C=O stretch, respectively. Diaryl or dialkyl ethers have a single intense band from the C—O stretch near 1300 or 1000 cm$^{-1}$, respectively, whereas an alkyl-aryl ether will have two bands, one near each end of the range from the two different C—O bonds (Figure 2.28).

Several functionalities that contain S=O groups give rise to bands in the region between 1300 and 1100 cm$^{-1}$ (Figures 2.29 on page 45 and 2.30 on page 46). In each, the bands are intense because of the large dipole moment of the S—O bond. The presence of sulfur can be verified by the mass spectrum to help in assigning bands in the region around 1200 cm$^{-1}$.

Finally, some chlorine-containing compounds can be identified by observation of the C—Cl stretch. Aromatic chlorides absorb around 1000 cm$^{-1}$, whereas aliphatic C—Cl stretches can sometimes be seen as strong bands between 800 and 600 cm$^{-1}$ (Figure 2.31, page 46). Again, you have to be careful in drawing conclusions from

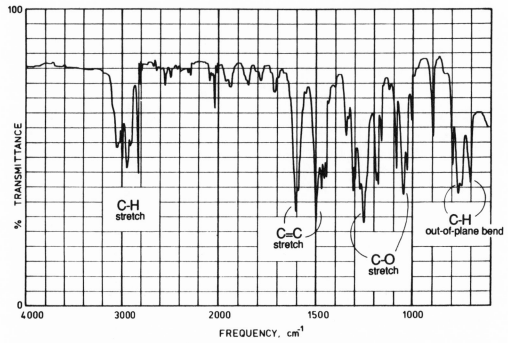

**Figure 2.28**
Infrared spectrum of anisole (thin film) showing the two C—O stretch bands.

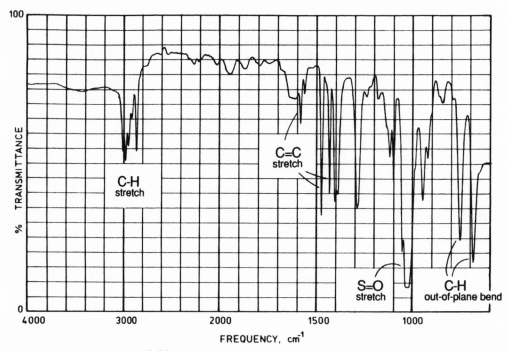

**Figure 2.29**
Infrared spectrum of benzyl methyl sulfoxide (thin film).

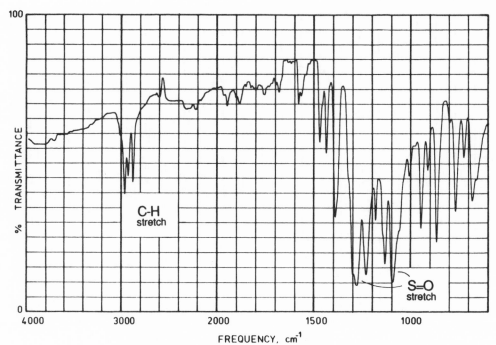

**Figure 2.30**
Infrared spectrum of benzyl methyl sulfone (KBr pellet).

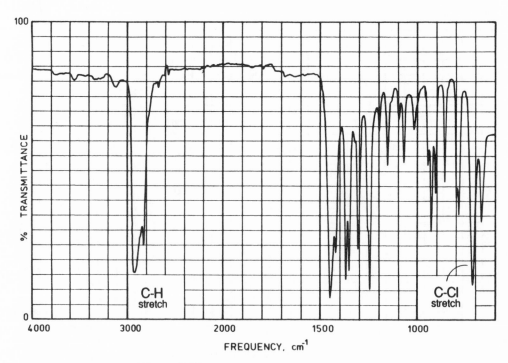

**Figure 2.31**
Infrared spectrum of neopentyl chloride (thin film).

C—Cl stretching vibrations, because other bands also appear at low frequencies as a result of C—C bending vibrations.

You may be able to identify a compound from peaks in the fingerprint region if the molecule is simple enough that a characteristic vibration is isolated. *Be careful not to overinterpret the spectrum.* Detailed information about the C—H framework, for example, is more easily obtained from the NMR spectrum.

A complete summary of absorption frequencies for many functional groups is collected in Figure 2.32 (on pages 48 and 49). You may sometimes find it helpful to see if any other peaks in the spectrum can be identified from the data in that figure.

## B. The Heteroatom Approach

The previous section outlined a strategy that provides a general survey of the IR spectrum without requiring additional spectroscopic information except for subsequent confirmation of assignments. However, as was discussed in Chapter 1, you can use the mass spectrum initially to find the molecular formula of the compound being studied (Figure 1.3). If you know the molecular composition of the unknown, you have an alternative strategy for approaching the interpretation of the IR spectrum, one that targets possible functional groups in terms of the heteroatoms present. The following sections are organized according to the types of atoms present besides carbon and hydrogen. You need to be aware that, in addition to the functional groups containing heteroatoms, C≡C bonds, C=C bonds, and aromatic rings may also be present, contributing peaks to the spectrum.

### 1. Compounds Containing Only Oxygen

If the molecular formula for the compound under investigation contains only C, H, and O, there are only a few possible functional groups. First, examine the region between 3600 and 3200 cm$^{-1}$. A broad strong band indicates an O—H group (Section III.A.3). Next examine the region between 1800 and 1650 cm$^{-1}$, where an intense band indicates the presence of a carbonyl group (Section III.A.1). If neither of the regions just mentioned has a readily noticeable band, then the molecule must be an ether, and you should check the region between 1300 and 1000 cm$^{-1}$ for the C—O stretch (Section III.A.6).

### 2. Molecules Containing Nitrogen

The presence of nitrogen can be readily detected by mass spectroscopy, especially if there are an odd number of nitrogen atoms in the molecule (Chapter 5). However, interpretation of the IR spectrum is sometimes difficult for certain types of amines.

You can begin by examining the region between 3500 and 3200 cm$^{-1}$. Broadened bands in this region signify that the molecule contains either an amino or amido group (Section III.A.3; remember that a sharp band at 3300 cm$^{-1}$ may be a ≡C—H

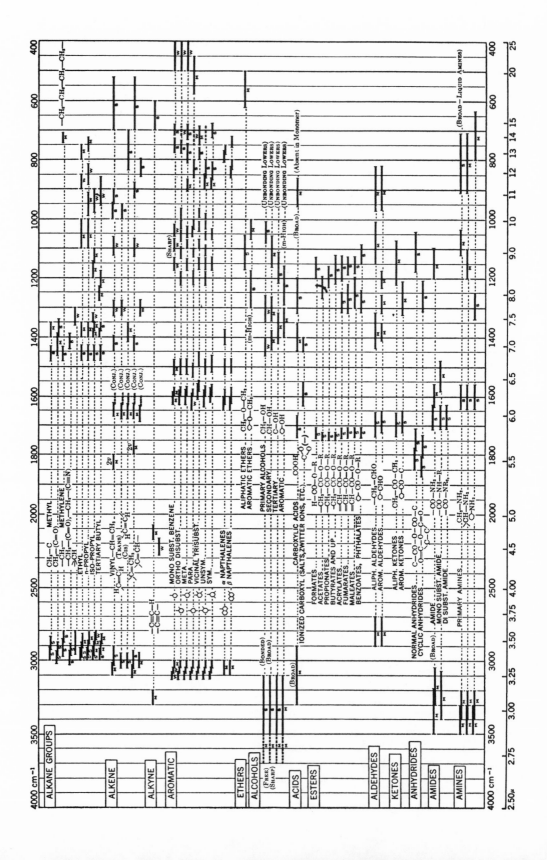

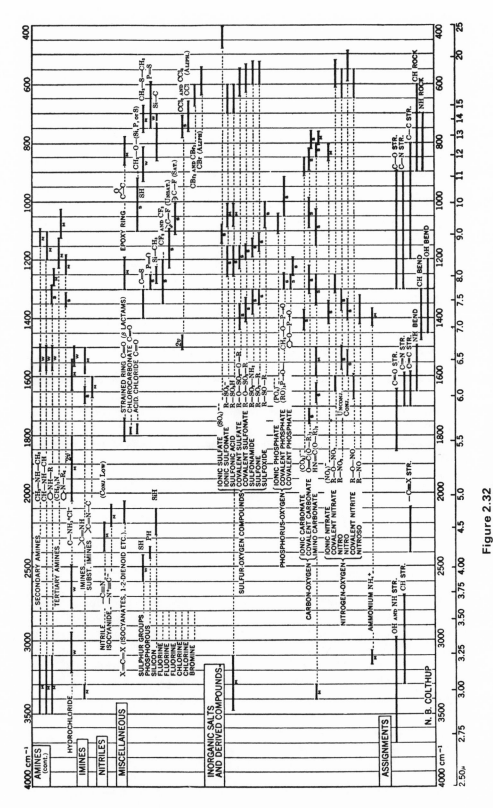

**Figure 2.32**
Colthup chart of characteristic group infrared absorptions.

49

stretch). To differentiate between an amine and an amide, check the carbonyl region, since an amide group has a characteristically low-frequency $C\!=\!O$ stretch (Section III.A.1).

If no bands are observed above 3000 cm$^{-1}$, check the region around 2200 cm$^{-1}$. This is where the $C\!\equiv\!N$ stretch occurs (Section III.A.2). If no peak is present there, look at the region around 1650 cm$^{-1}$ for the $C\!=\!N$ stretch (Section III.A.4). Remember, too, that $C\!=\!C$ vibrations appear around 1600$^{-1}$.

Finally, if no bands are observed in the $C\!=\!N$ stretch region, look for a single *strong band* between 1600 and 1500 cm$^{-1}$ (nitroso group) or for *two strong bands*, 1600–1500 cm$^{-1}$ and 1390–1300 cm$^{-1}$ (nitro group; Section III.A.4). Obviously oxygen must be in the molecular formula if you think an $N\!=\!O$ or $NO_2$ group is present.

If there are no bands in any of these regions, you may assume that the molecule is a tertiary amine. Unfortunately, the only bands that are diagnostic for a 3° amine are the $C\!-\!N$ stretches that occur in the region 1300–1000 cm$^{-1}$, analogous to $C\!-\!O$ stretches. Because a $C\!-\!N$ stretch is not as intense as a $C\!-\!O$ stretch, other peaks can interfere with identification of the $C\!-\!N$ stretching vibration.

### 3. Molecules Containing Sulfur

The presence of sulfur in a molecule is often readily detected by examination of the mass spectrum. Furthermore, many sulfur-containing compounds are foul-smelling substances, and you may suspect that the molecule under study has one or more sulfur atoms simply from its stench.

The first part of the IR spectrum to check is around 2500 cm$^{-1}$, where absorption from the $S\!-\!H$ stretch appears (Section III.A.2). If that region is blank, and the molecule also contains oxygen, then you should examine the region between 1400 and 1100 cm$^{-1}$, where absorptions from the $S\!=\!O$ stretch are observed (Section III.A.6). If no bands are observed in that region, or if the molecule contains no oxygen, then the compound is most likely a sulfide (thioether, $R\!-\!S\!-\!R'$), for which there are no diagnostic bands in the IR region,

### 4. Molecules Containing Halogens

The presence of the halides Cl, Br, and I is readily detected from the mass spectrum. However, bands in the IR region are usually at very low frequencies, out of the range of normal spectrometers. If chlorine is present, you can check the carbonyl region to see if the molecule is an acid chloride (Section III.A.1). If there is no carbonyl band, check the region 1100–1050 cm$^{-1}$, where absorptions from aromatic $C\!-\!Cl$ stretches occur, or 800–600 cm$^{-1}$, where bands from aliphatic $C\!-\!Cl$ stretches appear. Since the $C\!-\!Cl$ stretch gives rise to a strong band, you may observe the band readily if there are not many other absorptions in the region from 1000–600 cm$^{-1}$.

## References

For those interested in more detailed discussions of the theory of infrared spectroscopy or for additional correlation charts of IR absorption bands, the following texts are suggested.

J. R. Dyer, *Applications of Absorption Spectroscopy of Organic Compounds.* Englewood Cliffs, N.J.: Prentice-Hall, 1965.

K. Nakanishi, *Infrared Absorption Spectroscopy-Practical.* San Francisco, Calif.: Holden-Day, 1962.

D. L. Pavia, G. M. Lampman, and G. S. Kriz, Jr., *Introduction to Spectroscopy.* Philadelphia, Pa.: Saunders, 1977.

V. M. Parikh, *Absorption Spectroscopy of Organic Molecules.* Reading, Mass.: Addison-Wesley, 1974.

R. M. Silverstein, G. C. Bassler, and T. C. Morrill, *Spectrometric Identification of Organic Compounds.* New York: Wiley, 4th ed., 1978.

# CHAPTER 3

# Nuclear Magnetic
# Resonance Spectroscopy

Nuclear magnetic resonance (NMR) spectroscopy has become the most useful technique for identifying organic compounds, because nearly all organic molecules contain protons. Using this technique, you can identify the environment of any group within a molecule that contains hydrogen atoms; moreover, you can deduce the connectivity of the proton-containing fragments. That information, together with a knowledge of the functional groups present, is often all you need to identify the molecule.

## I. THEORY

### A. The Absorption Process

Atomic nuclei that contain an *odd* number of protons and neutrons display a magnetic moment caused by a quantized spin of each nuclear particle. The simplest nucleus is that of the hydrogen atom which consists of a single proton. The spin angular momentum of the proton's spin is associated with the spin quantum number, I, which is either $-\frac{1}{2}$ or $+\frac{1}{2}$. For isolated hydrogen atoms, there is no inherent advantage to having a clockwise or counterclockwise spin; so the relative energies are the same. By definition, the spins are said to be *degenerate*.

When placed in an external magnetic field, however, a proton's own magnetic field, caused by the nuclear spin, can align either with or against the external field (Figure 3.1). If the spin generates a local field in the same direction as the external field,

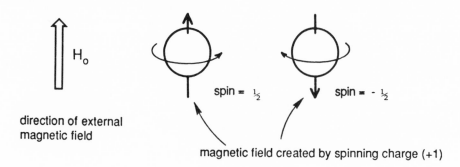

**Figure 3.1**
Alignment of the magnetic dipoles for a proton in a magnetic field.

then the spin quantum number is $+\frac{1}{2}$ and the spin is said to be "aligned." If the proton spin creates a local field in the opposite direction, the proton's spin quantum number is $-\frac{1}{2}$; and the spin is described as "opposed" to the field. The protons with nuclear spins aligned with the field are of lower energy than the protons with spins that are opposed; so a splitting of the energy degeneracy occurs (Figure 3.2).

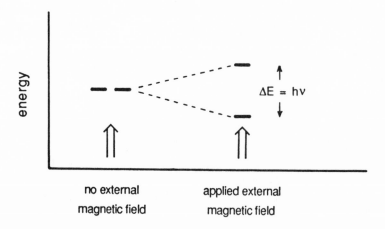

**Figure 3.2**
Splitting of energy levels for a hydrogen nucleus in a magnetic field.

If we look in more detail at the proton aligned with the magnetic field (spin $+\frac{1}{2}$), we would see the situation illustrated in Figure 3.3. Remember that the spinning of the proton creates a local magnetic field or "dipole"; and according to the laws of physics,

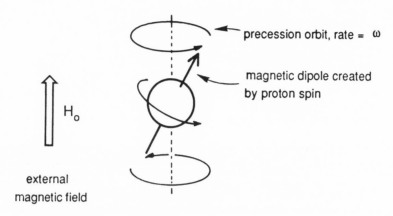

**Figure 3.3**
Precession of a proton's magnetic dipole in a magnetic field.

the field *precesses* about the axis of the external magnetic field, $H_0$. The rate of precession, $\omega$, depends on the strength of $H_0$, and a stronger external field causes a faster precession frequency for the proton.

The precession process also generates an electric field with a frequency $v$. If the sample is now irradiated with radiation in the radio-wave region (megahertz, MHz), the proton can absorb the necessary energy (Equation 3.1) and be promoted to the higher, less-favorable state (spin $-\frac{1}{2}$). This absorption is called "resonance," because the frequencies of the applied radiation and the precession coincide or resonate. The frequency necessary to cause the spin change, although related to energy by Equation 3.1, also correlates with field strength as given by Equation 3.2:

$$E = hv \qquad (v \text{ in MHz}), \tag{3.1}$$

$$v = \frac{\gamma H_0}{2\pi}. \tag{3.2}$$

The constant $\gamma$ is called the magnetogyric, or gyromagnetic, ratio, and is characteristic for each magnetically active nucleus. For the proton, $\gamma = 2.674 \times 10^4\,\text{gauss}^{-1}\text{sec}^{-1}$.

If we calculate the frequency that will resonate with a proton in various structures, we generate the values in Table 3.1 (on page 56), which relate $v$ to field strength, $H_0$. When you talk with chemists who work with NMR spectra, you will hear them mention

**Table 3.1**
Frequencies and field
strengths for proton resonance

| $H_0$ (Gauss) | $v$ (MHz) |
| --- | --- |
| 14,100 | 60 |
| 23,500 | 100 |
| 47,000 | 200 |
| 51,480 | 220 |
| 58,740 | 250 |
| 94,000 | 400 |
| 117,500 | 500 |

the numbers listed in the right-hand column of Table 3.1, because these numbers correspond to the common frequencies for proton NMR instruments. For example, the spectra in this text were recorded on a "200 MHz" NMR spectrometer; so you know approximately what field strength was required to record the spectra. An important parameter is the spectrometer frequency (hence field strength), which can change the appearance of the NMR spectrum. The effect of the frequency will be discussed later.

## B. Relaxation

The absorption process described in the preceding is straightforward; however, nuclei in the excited state must also be able to "relax" and return to the ground state. In IR spectroscopy, the vibrationally excited molecules relax by losing heat. In NMR spectroscopy, the nuclei return to the ground state by two processes, called "spin-lattice" and "spin-spin" relaxation. In the spin-lattice process, the energy is transferred to the molecular framework ("lattice") and is lost as translational or vibrational energy (heat). The half-life for this mechanism is referred to as $T_1$. Spin-spin relaxation occurs by transfer of the energy from one nucleus to neighboring nuclei, and the time required for this process is called $T_2$. Since the width of peaks in an NMR spectrum is inversely proportional to the time the molecule spends in the excited state, the magnitudes of $T_1$ and $T_2$ can be important, although they will not be covered further in this text. For most organic molecules in dilute solution, line widths are extremely narrow and thus well-resolved.

## C. Chemical Shifts

If all protons in a molecule absorbed radio waves of the same frequency, NMR spectroscopy would not exist. Fortunately, each proton absorbs at a slightly different frequency because of the influence of the surrounding electrons. If you have studied physics, you know that a magnetic field causes charged particles to move. Furthermore, you will recall that a moving charge creates its own magnetic field. In an organic molecule, the secondary field created by electrons in an external magnetic field *opposes*, and thus decreases, the applied field. The electrons are said to *shield* the nucleus, because the field experienced by the nucleus is slightly less than the actual applied field. If you look back at Equation 3.2, you will see that shielding (smaller $H_0$) will allow a lower frequency to cause resonance. The slightly different frequency absorptions give rise to many of the peaks seen in an NMR spectrum.

How small are the frequency differences? Consider the compound $CH_3OCH_2Br$, the spectrum of which is shown in Figure 3.4.

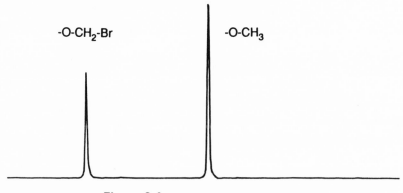

**Figure 3.4**
Proton NMR spectrum of $CH_3OCH_2Br$.

We can record the spectrum by two methods. In the first, we will use a fixed external field $H_0 = 47,000$ gauss. If the $CH_3$ group requires a frequency of 200,000,000 Hz (200 MHz) to cause resonance, the $CH_2$ will require a frequency of 200,000,440 Hz. By the second method, we will use a fixed frequency of 200 MHz and vary the field. If the $CH_3$ group requires an external field of 47,000 gauss to cause resonance, then the $CH_2$ group will require a field of 47,000.1034 gauss. Because of the difficulty in measuring either the field strength or the frequency with such precision, we need a relative measurement that does not require distinguishing small differences between large numbers. Thus, the positions of peaks in an NMR spectrum are presented in *parts per million* (ppm), which requires that neither frequency nor field strength be specified. Notice that for each method, the methylene group resonance differs by 2.2 ppm from the resonance for the methyl group:

$$\frac{(200.00044 - 200)\,\text{MHz}}{200\,\text{MHz}} = 2.2 \times 10^{-6} = 2.2\ \text{ppm},$$

or

$$\frac{(47,000 - 47,000.1034)\,\text{gauss}}{47,000\,\text{gauss}} = 2.2 \times 10^{-6} = 2.2\ \text{ppm}.$$

One consequence of using ppm to define the resonance position is that the NMR spectrometer can scan frequency with a fixed field or it can scan field strength with a fixed frequency. However, the resonance positions must be defined relative to the resonance signal of an internal standard. The most commonly used standard is tetramethylsilane, $(CH_3)_4Si$ (TMS), because most protons in organic compounds are less shielded than the protons in TMS. The position of the absorption relative to TMS is called the *chemical shift*: its designation is "$\delta$" and its units are "ppm."

Now consider the spectrum of $CH_3OCH_2Br$ again, this time as displayed in Figure 3.5. The peak for TMS is defined as $\delta = 0$ ppm. The absorptions for the methyl and

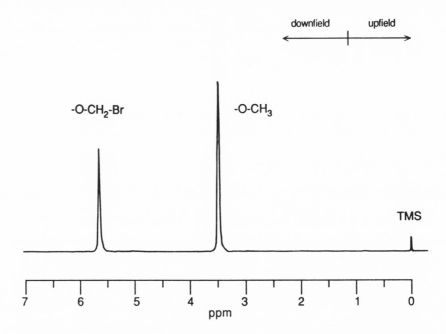

**Figure 3.5**
Proton NMR spectrum of $CH_3OCH_2Br$ with an internal standard of tetramethylsilane (TMS), showing the chemical-shift values in ppm.

methylene groups appear *downfield* from TMS at 3.5 and 5.7 ppm, respectively. The terms "downfield" and "upfield" are relative; the signal for the methyl group is upfield from the signal for the methylene group, but downfield of the TMS signal.

Figure 3.6 shows the *general* chemical-shift ranges for protons in common organic molecules. It is a good idea to learn the broad ranges, because much can be learned about a molecule just from a glance at the NMR spectrum. For example, if the IR spectrum of a compound indicates the presence of a carbonyl group, a quick check of

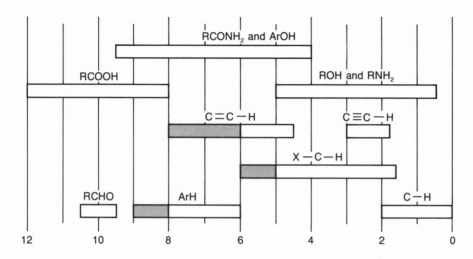

**Figure 3.6**
General chemical-shift ranges for protons. The crosshatched regions indicate extended ranges when additional electron-withdrawing groups are attached.

the NMR spectrum in the region 9.5 to 10.5 ppm would confirm the presence (or absence) of an aldehyde group. More detailed correlation charts will be presented in Section IV.

The different ranges for protons arise from several electronic phenomena within the molecular structure, including:

(1)  electronegativity of neighboring groups or atoms;

(2)  hybridization;

(3)  acidity and hydrogen-bonding; and

(4)  magnetic anisotropy.

(1) *Electronegativity*   effects are perhaps the easiest to understand. It should be apparent that if electron density is withdrawn from around the hydrogen atom nucleus toward a more electronegative atom, the proton will be deshielded and will absorb at a position farther downfield:

$$
\begin{array}{ccc}
\underset{\overset{|}{\diagdown}}{\text{H}} & & \underset{\overset{|}{\diagdown}}{\text{H}} \\
\overset{}{\text{C}} - \text{C} & vs. & \overset{}{\text{C}} \longrightarrow \text{X}
\end{array}
\qquad (3.3)
$$

For example,

$$CH_3 \longrightarrow CH_3 \qquad CH_3 \longrightarrow Cl \qquad CH_3 \longrightarrow OCH_3 \qquad (3.4)$$

δ 0.26                    δ 3.06                    δ 3.24

(2) *Hybridization*   effects are also easy to understand, because an s-orbital does not extend as far as a p-orbital; so increased s-orbital contribution to the C—H bond will result in less electron density in the bond, hence deshielding. However, hybridization effects are sometimes outshadowed by the influence of magnetic anisotropy, as we will see, and so become relatively unimportant.

(3) *Acidity* and *hydrogen-bonding*   affect the chemical shift by causing deshielding of the protons. Acidic protons are bound to heteroatoms, usually oxygen or nitrogen, and the positions of their resonances depend on temperature, solvent, and structure. As an extreme example, we can consider an acidic proton floating "free" in solution, the ultimate deshielded proton. In reality, of course, such protons are solvated or rapidly exchanging between molecules, and so are always bound to some atom. Exchange with more deshielded sites will cause the resonance to appear far downfield. Note that carboxylic acids absorb very far downfield from TMS:

$$CH_3\text{-}CH_3\text{-}O \longrightarrow H \qquad\qquad CH_3\text{-}\overset{\overset{\displaystyle O}{\|}}{C}\text{-}O \longrightarrow H \qquad (3.5)$$

δ 4.5                                      δ 10-13

(4) *Magnetic anisotropy*   has one of the greatest influences on resonance frequencies. The phenomenon arises from unsaturation, in which the electrons in the π-bonds create nonuniform fields in the vicinity of protons. For some protons, the effect of these fields will be shielding; for others, the effect will be deshielding. Aromatic compounds provide the classic case of magnetic anisotropy as illustrated for benzene in Figure 3.7.

When benzene is placed in a magnetic field, the delocalized electrons of the π-system circulate in a direction that creates a secondary field passing through the center of the ring in a direction opposite to the external field. However, at the proton positions on the periphery of the ring, we can see that the secondary field actually operates in the *same* direction as the applied field. The result is that the protons on benzene experience a slightly greater field than they would in the absence of the π-electrons; so a

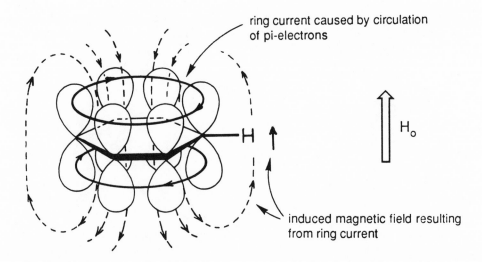

ring current caused by circulation
of pi-electrons

$H_o$

induced magnetic field resulting
from ring current

**Figure 3.7**
Magnetic anisotropy in benzene, showing the effect of the local magnetic field on the proton
attached to the ring.

slightly higher frequency is needed to cause resonance. The protons of benzene and
other aromatic compounds therefore appear farther downfield from TMS than would
be expected simply from hybridization or electronegativity effects. Protons located
directly above the ring will be *more* shielded, and many compounds have been prepared
to test that hypothesis. The cyclophane shown in Figure 3.8 is a good example,
illustrating how shielded such protons can be.

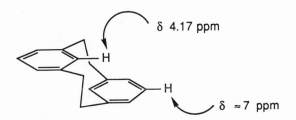

δ 4.17 ppm

δ ≈7 ppm

**Figure 3.8**
Effect of magnetic anisotropy on the chemical shifts of protons
in [2.2]-metacyclophane. The proton that appears farther upfield
is located directly within the pi-cloud of the neighboring
benzene ring.

The $\pi$-bonds of other types of compounds also can create local magnetic fields (see Figure 3.9). As an example of the shifts caused by anisotropy, compare the resonance

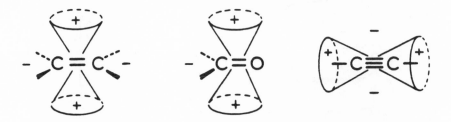

**Figure 3.9**
Magnetic anisotropy effects in other types of unsaturated organic compounds. The ($+$) sign represents shielding effects, and the ($-$) sign deshielding effects.

positions for the protons in the two compounds in (3.6). Each substituent connected to the double bond is shifted downfield relative to the analogous group in the saturated compound.

$$
\begin{matrix}
\text{H} & & \text{CH}_3 \\
\diagdown & & \diagup \\
& \text{C}=\text{C} & \\
\diagup & & \diagdown \\
\text{H} & & \text{CH}_3
\end{matrix}
\qquad\qquad
\begin{matrix}
\text{H} & & \text{CH}_3 \\
\diagdown & & \diagup \\
\text{H}-\text{C}-\text{C}-\text{H} & \\
\diagup & & \diagdown \\
\text{H} & & \text{CH}_3
\end{matrix}
\tag{3.6}
$$

$\delta$ 4.75   $\delta$ 1.70        $\delta$ 0.89   $\delta$ 0.89

The protons of acetylene present an especially interesting situation. You might predict that the proton resonances would be downfield from those of alkenes because of hybridization effects (i.e., more s-character in the C—H bond). However, the resonances actually appear upfield from those of ethylene, because of magnetic anisotropy effects:

$$
\text{H}-\text{C}\equiv\text{C}-\text{H}
\qquad\qquad
\begin{matrix}
\text{H} & & \text{H} \\
\diagdown & & \diagup \\
& \text{C}=\text{C} & \\
\diagup & & \diagdown \\
\text{H} & & \text{H}
\end{matrix}
\tag{3.7}
$$

$\delta$ 2.01                        $\delta$ 5.29

Another special situation is the aldehyde proton, which appears very far down-field. Again, magnetic anisotropy, coupled with the electronegativity of the carbonyl oxygen atom, causes a downfield shift in the aldehyde proton resonance:

$$\delta\ 5.7 \qquad\qquad \delta\ 9.9 \qquad\qquad \delta\ 3.6 \qquad (3.8)$$

## D. Chemical Equivalence

As you saw in $CH_3-O-CH_2Br$, there are five protons, but only two signals. The protons of the methyl group and those of the methylene group are said to be "chemically equivalent"; hence their chemical shifts are identical. To decide whether a proton is chemically equivalent to others, simply consider replacing each proton separately by another group. Replacements that generate *identical* compounds define chemically equivalent protons. For a methyl group, replacement of any of the three protons gives the same product; therefore, those three protons are chemically equivalent and their resonance signals appear at the same chemical shift value (Figure 3.10, page 64).

Protons bound to the same carbon atom are generally equivalent; however, if there are conformations that prohibit free rotation about certain bonds within a molecule, then the protons will not necessarily be equivalent. Nonequivalence of protons bound to the same carbon atom occurs in cyclic compounds, or in acyclic compounds at low temperatures, where rotation about carbon-carbon bonds is slowed substantially. Some examples are:

$$(3.9)$$

(at low temperature)

The area of the resonance signal for chemically equivalent protons is proportional to their number, a point to which we will return in Section IV.

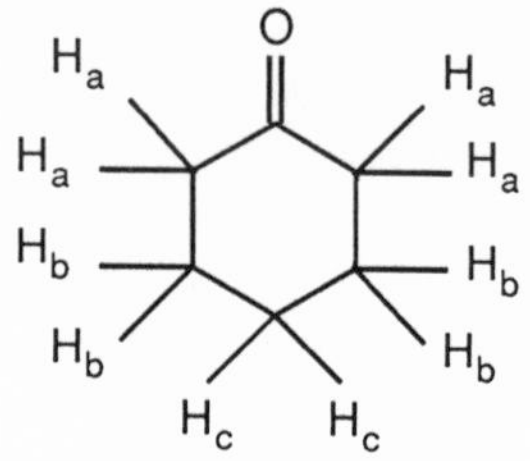

| | |
|---|---|
| $H_a$ | 2.2 |
| $H_b$ | 1.8 |
| $H_c$ | 1.7 |

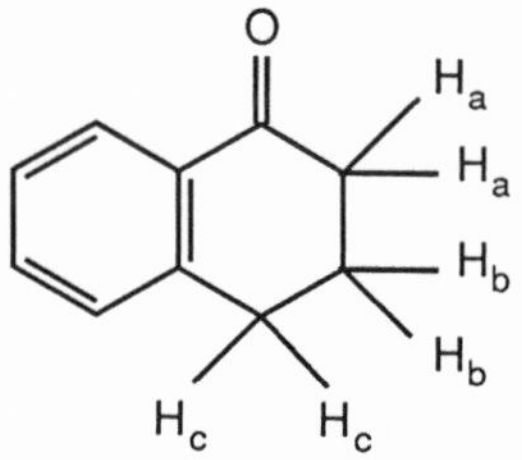

| | |
|---|---|
| $H_a$ | 2.58 |
| $H_b$ | 2.18 |
| $H_c$ | 2.92 |

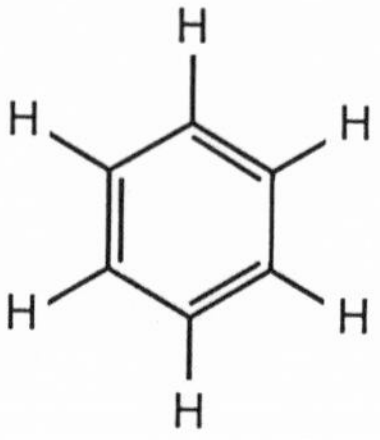

all protons at 7.26

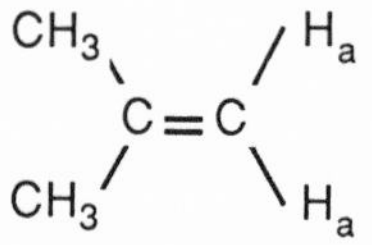

| | |
|---|---|
| $CH_3$ | 1.70 |
| $H_a$ | 4.75 |

**Figure 3.10**

Examples of chemical-shift values of chemically equivalent hydrogen nuclei.

## E. Spin-spin Splitting

The chemical shift by itself would make NMR spectra very useful for discovering the structure of organic compounds; but, in fact, much more can be learned about a molecule's structure from the NMR spectrum. If you were to predict the appearance of the NMR spectrum for $\alpha$-chloroacetone ($Cl-CH_2-CO-CH_3$) using correlation charts or your knowledge about relative electronegativities, you would expect to see two signals, one at about 4.2 ppm and one at about 2.3 ppm. The actual spectrum, shown in Figure 3.11, confirms this expectation.

On the other hand, if you had to predict the spectrum of propanoyl chloride, $CH_3CH_2COCl$, you might again expect to see two signals, one at about 2.9 ppm for the methylene group, and another at about 1.2 ppm for the methyl group. The actual spectrum, shown in Figure 3.12, clearly contains many more peaks than you would predict from knowledge of only the chemical shifts. Why?

The situation arises because the nuclei of protons on the carbon *adjacent* to the one being examined have different alignments in the magnetic field. Look first at the

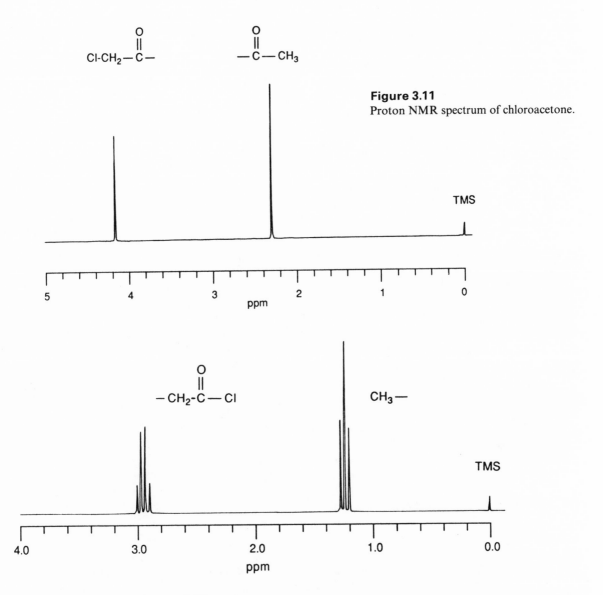

**Figure 3.11**
Proton NMR spectrum of chloroacetone.

**Figure 3.12**
Proton NMR spectrum of propanoyl chloride.

protons on the methyl group. All three protons are chemically equivalent; so only a single resonance should be observed when the compound is placed in magnetic field. However, the protons on the adjacent methylene group affect the local magnetic field experienced by the methyl-group protons. Remember that their nuclear spins can be aligned either with or against the external magnetic field; so in each molecule, the methylene protons can have one of the spin combinations ($\uparrow$ = spin $\frac{1}{2}$ and $\downarrow$ = spin $-\frac{1}{2}$) shown in (3.10):

$$CH_3 - CH_2 - COCl \qquad CH_3 - CH_2 - COCl$$

(a) $\qquad \uparrow\uparrow$ $\qquad\qquad$ (b) $\qquad \uparrow\downarrow$

$$CH_3 - CH_2 - COCl \qquad CH_3 - CH_2 - COCl$$

(c) $\qquad \downarrow\uparrow$ $\qquad\qquad$ (d) $\qquad \downarrow\downarrow$

(3.10)

In molecule (a), the methyl group experiences a stronger apparent field than it would if the methylene group were absent, because the neighboring spins create small fields that *add* to the external field. The methyl group in this molecule will have a resonance frequency slightly downfield from that for an unperturbed methyl group in the same environment. In molecules (b) and (c), the influence of the tiny magnetic fields resulting from the nuclear spins of the methylene group is zero, because the fields cancel. The combinations in (b) and (c) are actually the same energetically, since there is no way to label the two protons differently (i.e., they are chemically and magnetically equivalent); so the methyl resonances for molecules (b) and (c) will occur in the same place as resonances for an unperturbed methyl group. Finally, the methyl group in (d) will experience a slightly lower field strength, because the fields generated by the methylene protons will counteract the external field. Its resonance will occur at a slightly lower frequency than that for an unperturbed methyl group, and so will appear farther upfield.

If we add the four peaks together, we will generate a pattern (Figure 3.13) corresponding to the pattern observed for the methyl signal in the actual spectrum (Figure 3.12).

You can go through the same procedure for the methylene protons influenced by the small magnetic fields created by the three methyl protons. For a $CH_2$ group adjacent to a methyl group, there will be four peaks, created by the spin orientations of the methyl protons shown in (3.11):

$$\uparrow\uparrow\uparrow \qquad \uparrow\uparrow\downarrow \qquad \uparrow\downarrow\uparrow \qquad \downarrow\uparrow\uparrow \qquad \uparrow\downarrow\downarrow \qquad \downarrow\uparrow\downarrow \qquad \downarrow\downarrow\uparrow \qquad \downarrow\downarrow\downarrow \qquad (3.11)$$

When grouped with the molecules of comparable energy, the four spin orientations generate the set of local fields shown in (3.12):

$$
\begin{array}{cccc}
 & \downarrow\uparrow\uparrow & \downarrow\downarrow\uparrow & \\
 & \uparrow\downarrow\uparrow & \downarrow\uparrow\downarrow & \\
\uparrow\uparrow\uparrow & \uparrow\uparrow\downarrow & \uparrow\downarrow\downarrow & \downarrow\downarrow\downarrow
\end{array}
\qquad = \qquad
$$

(3.12)

Thus, a methylene group next to a methyl group should be split into a quartet with peak ratios of $1:3:3:1$, as is observed for propanoyl chloride (Figure 3.12).

Proton resonances will always be influenced through the intervening bonds by the local fields generated by hydrogen nuclei on adjacent atoms. The effect is termed "spin-

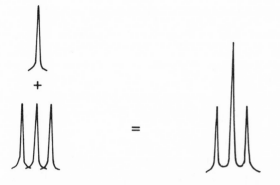

**Figure 3.13**
Additive effects of magnetic fields from neighboring protons to generate the observed triplet pattern for the methyl signal in propanoyl chloride.

spin splitting" (or coupling) and can be summarized by several generalizations, as follows.

(1) *The "N + 1 Rule."* The resonance for a set of protons will be split into $N + 1$ peaks, where $N$ = the number of equivalent protons on the *adjacent* carbon atoms. The $N + 1$ rule holds even when the protons are distributed over more than one carbon atom, as shown in Figure 3.14 for 2-chloropropane.

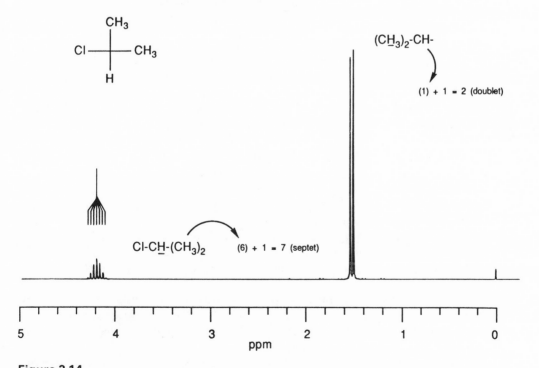

**Figure 3.14**
Proton NMR spectrum for 2-chloropropane, showing spin-spin splitting patterns. The underlined hydrogen atoms are the ones giving rise to the signal *before* splitting by the adjacent protons.

(2) *Effect of distance.*   Peak splitting is generally not observed among chemically equivalent protons, and the influence of neighboring protons will diminish, so that only hydrogen atoms within three bonds (i.e., H—C—C—H) will have much effect. This type of influence is called "vicinal coupling." Sometimes you will see coupling of protons on the same carbon atom (only two bonds apart), called "geminal coupling," especially for terminal alkenes in which the two protons are inequivalent. Unsaturation can promote splitting over longer distances, for example, in benzene rings (see Section IV).

(3) *Appearance.*   The ideal height ratios for the splitting pattern will follow the arrangement described by Pascal's triangle, shown in Figure 3.15. Thus the peaks in a

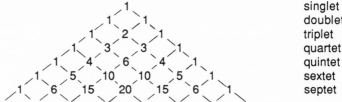

singlet
doublet
triplet
quartet
quintet
sextet
septet

**Figure 3.15**
Pascal's triangle: predicted intensities of peaks in patterns caused by spin-spin splitting.

doublet will have equal intensities; the peaks in a triplet will have a $1:2:1$ intensity ratio; a quartet will have a $1:3:3:1$ intensity pattern; etc.

(4) *Coupling constant.*   The magnitude of the splitting among protons (in Hz) is called "$J_{H-H}$", the coupling constant, and is characteristic of the type of protons causing the splitting. Figure 3.16 summarizes several types of coupling constants. Information about stereochemistry, including aryl and alkenyl substitution, can be obtained from knowing $J$ (see Section IV).

The coupling constant, unlike the chemical shift, does *not* vary with field strength. Thus the splitting patterns in spectra recorded on high-field NMR instruments will appear much more compact, even though the magnitude of $J$ is the same (Figure 3.17, page 70).

Finally, if one feature is split into a pattern with a coupling constant $J = x$ Hz, then there must be at least one other resonance split by $J = x$ Hz. If there appears to be only one feature with a particular $J$-value, you may be looking at coupling of the proton resonance with another NMR-active nucleus (e.g., $^{19}F$, $^{31}P$, etc.). Coupling to a different nucleus can be detected by recording the NMR spectrum at a different frequency, at which the other nucleus is in resonance.

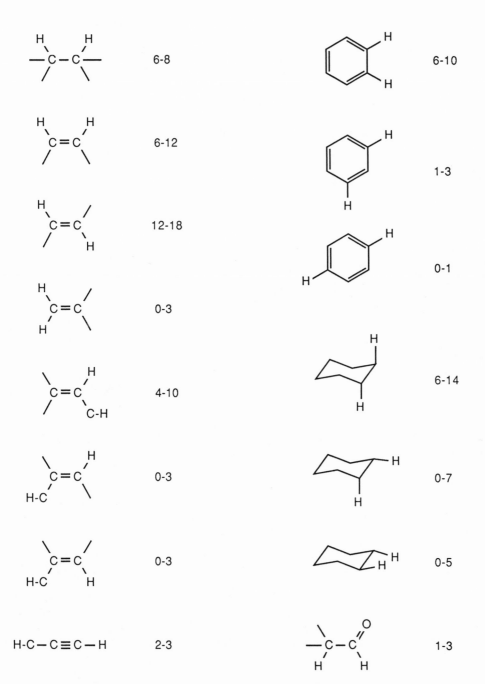

**Figure 3.16**
Summary of *J*-values for systems displaying spin-spin splitting (all values in Hz).

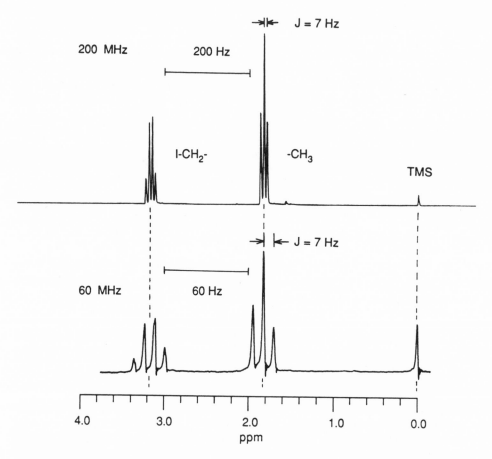

**Figure 3.17**

Appearance of the proton NMR spectrum of ethyl iodide at two different spectrometer frequencies. The coupling constant, $J$, is the same for each spectrum, but it appears narrower in the upper spectrum because the frequency scale is much wider.

(5) *Exchangeable protons.* Protons bonded to oxygen atoms in alcohols and nitrogen atoms in amines sometimes do not engage in spin-spin splitting, because they can undergo intermolecular exchange faster than the NMR spectrometer can record the $S = +\frac{1}{2}$ to $S = -\frac{1}{2}$ transition. In essence, they "appear" to the spectrometer to be not actually bound to the molecule. The rate of exchange is often influenced by impurities in the sample, especially water or acids; so the manner in which the sample is prepared can affect whether you will see splittings among protons bonded to O or N atoms.

## F. First-order Spectra

The spin-spin splitting interactions presented in the preceding occur only in situations that are considered "first-order." A first-order splitting pattern is obtained when the value of $J$ is much smaller than the difference in chemical shifts ($\Delta v$) for the two resonances that are interacting. When $\Delta v$ is measured in Hz (note: Hz, *not* ppm), splittings with a value of $\Delta v/J > 5$ can be considered first-order, and are easily interpreted because they follow the $N + 1$ rule. You should now be able to see why high-field NMR spectrometers are so useful: since stronger fields do not affect $J$, but do influence $v$, spectra run in high fields are more likely to be first-order, because the ratio $\Delta v/J$ is larger (*cf.* Figure 3.17). Fortunately, many spectra of complex molecules appear first-order, and several will be considered further in Section IV.

## G. Complex Spectra

A treatment of complex spectra is beyond the scope of this text, but a simple example will illustrate what sort of change is observed when we go from a first-order spectrum to a more complex one. Consider a series of molecules of the form $X_2CH-CHX'_2$. We would expect to observe two doublets in the NMR spectrum when X and X' have a large electronegativity difference, as illustrated in Figure 3.18a (on page 72). Note that even here the doublets "lean" toward each other (indicated by the dotted line); in fact, for complicated spectra, the leaning is a good clue about which group is causing the splitting of a resonance. As the electronegativity difference between X and X' becomes less pronounced, the chemical-shift difference, $\Delta v$, becomes smaller, approaching the value of $J$ (Figure 3.18b–e). For the spectra in this set of compounds, the inner peaks of the doublet grow in intensity, but the outer ones decrease. When X = X', the hydrogen atoms are chemically equivalent and their resonances appear, unsplit, at the same chemical-shift value (Figure 3.18f).

When you start to consider situations in which there is more than one hydrogen atom on two neighboring carbon atoms, the treatment of complex spectra becomes quite complicated, and will not be considered here. (If you are interested, consult the references at the end of the chapter; the text by Silverstein, Bassler, and Morrill gives an especially useful overview of the subject.)

## II. EXPERIMENTAL CONSIDERATIONS

Modern NMR spectrometers are sophisticated, computer-controlled instruments capable of measuring spectra of extremely small quantities. However, the basic components have not changed over the years; and they include a magnet, radio-frequency oscillator, detector, and recorder. The magnets on older and smaller instruments (60 to

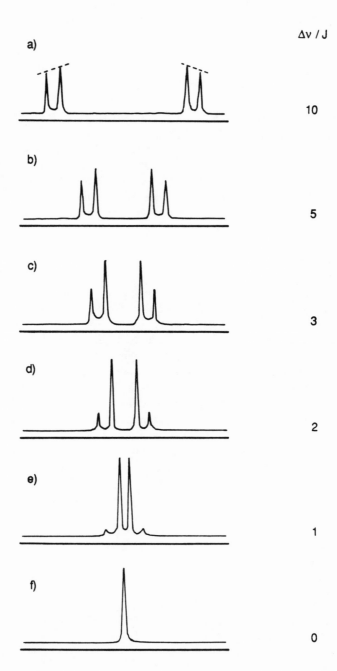

**Figure 3.18**
Effect of the ratio of $\Delta v/J$ on the appearance of two doublets in molecules
of the type $X_2CH-CHX'_2$.

100 MHz) are electromagnets, but high-field spectrometers (200 to 600 MHz) have magnets of the superconducting type.

Normally, the spectrum is recorded over the range from 10 to 0 ppm with the absorptions recorded as peaks on the "baseline" running near the bottom of the chart. Although you have already seen several spectra in the last section, Figure 3.19 presents a typical NMR spectrum, summarizing some of the relationships discussed above. Integrated intensities of the peaks, corresponding to the area under each of the absorptions, are plotted above each peak with a separate line. The importance of the integration curve, as it is called, is discussed in Section III.

The three features of the NMR spectrum you will use for interpretation are:

(1) the position of the absorption (in ppm);
(2) the number of peaks for each absorption (singlet, doublet, triplet, etc.) and the $J_{H-H}$ value; and
(3) the integrated intensity of each feature.

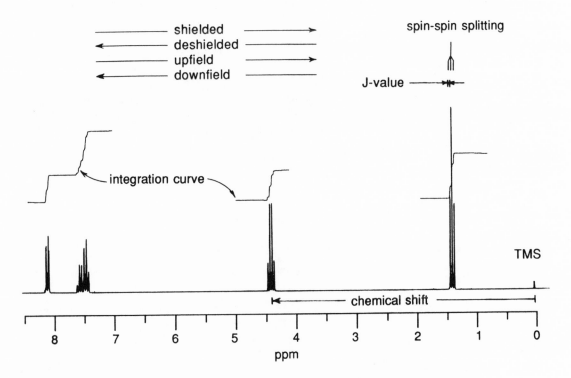

**Figure 3.19**
Appearance of a typical proton NMR spectrum showing a summary of terms used to discuss spectral features.

In IR spectra, only the position of the absorption is quantitatively important; but in NMR spectroscopy, all three of these characteristics are quantitatively valuable.

To prepare a sample for an NMR spectrum recorded on a conventional instrument, you dissolve 10 to 50 mg of the compound in about 0.5 mL of solvent that contains a small amount of the standard, TMS, tetramethylsilane, $(CH_3)_4Si$. The resulting solution is introduced into a 5-mm, thin-walled glass tube that is placed in the spectrometer probe. The tube must be precision-made, so that it spins freely in the sample holder. Spinning the sample ensures that it experiences as homogeneous a magnetic field as possible. The more sensitive, high-field spectrometers use a Fourier-transform technique for recording the spectrum; and samples on the order of $\mu g$ can be routinely examined, because hundreds of scans are recorded and stored by the computer. (If you are interested, a more complete description of FT NMR spectrometers can be found in Farrar's text, listed at the end of the chapter.)

The most important characteristic of the solvent is that it not contain any protons, so that extra resonances will not be present to interfere with those of the compound being studied. Carbon tetrachloride seems ideal, although it has two disadvantages: first, many organic compounds are not particularly soluble in it; second, it is carcinogenic. Deuteriochloroform, $CDCl_3$, is the most widely used solvent, followed by perdeuterioacetone, $CD_3-CO-CD_3$, perdeuteriomethanol, $CD_3OD$, perdeuteriobenzene, $C_6D_6$, and deuterium oxide, $D_2O$. Deuterium does have a nuclear spin, but its resonance frequency is quite different from the frequency used to obtain proton spectra, and so does not interfere. Actually, small peaks will still be apparent in the proton spectrum because the isotopic purity of deuteriated solvents is usually only 98 to 99.8 percent. Table 3.2 summarizes the positions at which residual peaks from nondeuteriated solvents are commonly observed.

**Table 3.2**

Chemical shifts for common NMR solvents

| Solvent | | $\delta$ (ppm) |
|---|---|---|
| aetone-d$_6$ | $CD_3-CO-CD_3$ | 2.05 |
| benzene-d$_6$ | $C_6D_6$ | 7.20 |
| chloroform-d | $CDCl_3$ | 7.25 |
| deuterium oxide | $D_2O$ | ~5 (variable) |
| DMSO-d$_6$ | $CD_3-SO-CD_3$ | 2.5 |
| methanol-d$_4$ | $CD_3-OD$ | ~4.8 (variable) |
| | | 3.35 |
| methylene chloride-d$_2$ | $CD_2Cl_2$ | 5.35 |
| tetrahydrofuran-d$_8$ | $(CD_2-CD_2)_2O$ | 3.6 |
| | | 1.75 |

# III. INTERPRETATION OF THE NMR SPECTRUM

Even more so than for IR spectroscopy, interpretation of the NMR spectrum follows no single prescribed strategy, because the number of protons, their chemical shifts, and the spin-spin splitting patterns are so interrelated. However, the strategy presented here is one that will often work, and you can vary it as you acquire experience.

The first step is to convert the integration curves into the ratios for the number of protons associated with each feature. In this text a feature is defined as a single peak or group of peaks associated with one set of chemically equivalent protons. The second step is to assign the type of proton to each resonance feature according to the chemical shift. The final step is to assemble the fragments by analysis of the spin-spin splitting patterns.

## A. Integrated Intensity

If you examine a typical NMR spectrum (Figure 3.20), you will notice a separate line, traced above the actual spectrum, called the "integration curve." The height of the line above each feature is proportional to the area under the curve, hence to the number of

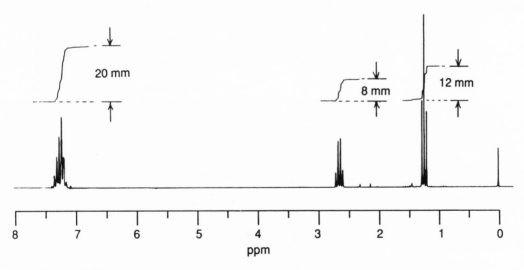

**Figure 3.20**
Proton NMR spectrum of ethylbenzene showing the integration curve and how it is measured.

protons giving rise to the peaks. For the spectrum in Figure 3.20, we can measure the height of the curve for each feature and obtain the values given. The lowest common factor for $20:8:12$ is $5:2:3$, which means there are at least $5 + 2 + 3 = 10$ protons in this molecule. Remember: integrated intensities give only ratios; so the total number of protons may be a multiple of that value (in this example, 10, 20, 30, etc.).

You will often be able to deduce the absolute number of protons simply from the appearance of the NMR spectrum, as we will see. Even better, from analysis of the mass spectrum, you can deduce the molecular formula of the compound, which will tell you the total number of protons.

## B. Chemical Shift

The chemical shift of each resonance depends on the local electronic environment about the proton; however, each feature can be assigned by comparison with empirical data collected for thousands of compounds. Figure 3.6 showed a schematic summary of principal chemical shift ranges, and Figure 3.21 gives a more detailed compilation.

As you saw for IR spectroscopy, assignments of absorptions in regions that are frequently blank are the least ambiguous and therefore easily made. Unlike the peaks of an IR spectrum, many of which cannot be assigned to specific fragments of the molecule, the peaks of an NMR spectrum should all be assigned, although some ambiguity may be unavoidable.

The logical place to begin is the region farthest downfield from TMS (9–12 ppm), where a peak can indicate only a few compound types. Alternatively, the region closest to TMS (0–1 ppm) is a good one to examine, since peaks in that region are rarely associated with anything but methyl groups.

### 1. Acids and Aldehydes: 12–9 ppm

The region between 12 and 9 ppm is where carboxylic acid and aldehyde protons appear, although for different reasons (see Section I). Acidic compounds can be easily differentiated from aldehydes, because protons associated with a heteroatom will often exchange readily with $D_2O$ (Equation 3.13).

$$RCOOH \ + \ D_2O \ \rightleftharpoons RCOOD + \ DOH$$
$$\delta \approx 11 \text{ ppm} \quad \text{no signal} \quad \text{no signal} \quad \delta \approx 5 \text{ ppm} \tag{3.13}$$

If we use an excess of $D_2O$, the equilibrium will be driven to the right, and the acid proton signal will disappear. Since the hydrogen atom of an aldehyde group is attached to carbon, it does not exchange with $D_2O$. The exchange technique is useful for many other functional groups in which the proton is attached to a heteroatom, especially oxygen. Alcohols, phenols, and carboxylic acids undergo exchange readily; amides and

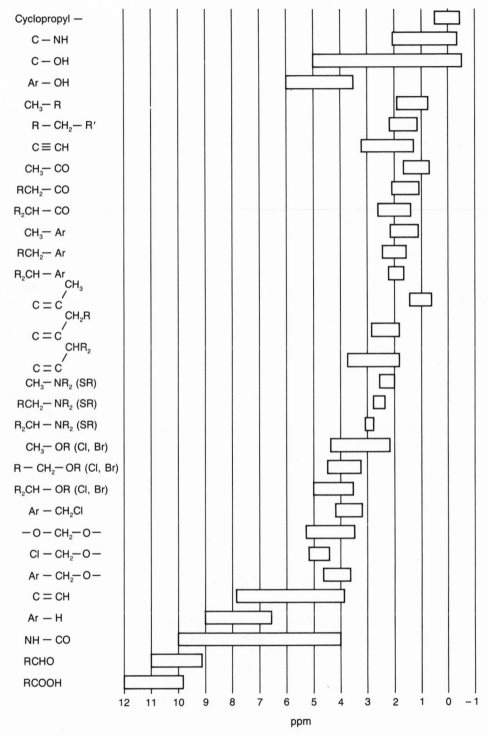

**Figure 3.21**
A detailed compilation of chemical shift ranges for proton NMR spectra.

thiols often need to be treated with a stronger base like NaOD in $D_2O$ to promote the exchange process. Although a polar solvent is preferred for carrying out the exchange reaction, compounds with exchangeable protons dissolved in $CDCl_3$ often react with $D_2O$ when a drop of the latter is added to the sample. The organic phase, containing the deuterium-exchanged compound, can then be carefully transferred to the NMR tube, leaving the $D_2O$ and DOH behind. Because DOH is slightly soluble in most organic solvents, its NMR signal will still be observed and must be taken into consideration.

### 2. Arenes and Alkenes: 8–5 ppm

The next region to examine is the one in which compounds having $sp^2$ C—H fragments have their resonances. Compounds in this category include arenes and alkenes. You can usually distinguish between these two types of compounds simply on the basis of the chemical shift (Figure 3.6). Unless an alkene has strongly electron-withdrawing groups attached to the double bond, the alkene protons will absorb between 5 and 6 ppm. In addition, the integrated intensity for the features in the region from 5 to 8 ppm may also be helpful, since a double bond (except for ethylene) can have at most three protons, whereas a phenyl ring will often have four or five protons. You can sometimes use the intensities of the peaks in the region 5 to 8 ppm to discover the absolute number of protons for the molecule under investigation, since you may have a good idea about the substitution pattern of the ring or double bond from analysis of the IR spectrum. For example, if the overtone region of the IR spectrum suggests a *meta*-disubstituted benzene compound, then you would know that the height of the integration curve for the aromatic region must correspond to four protons. Once you know the absolute number of protons for any one feature, then you can easily calculate how many hydrogen atoms are associated with each of the other peaks in the spectrum.

Although an analysis of the IR spectrum should be sufficient to define the substitution pattern of a benzene ring or double bond, a detailed analysis of the spin-spin splitting patterns for aromatic or alkenyl protons can confirm the conclusion. This aspect of the analysis will be covered in Section IV.

### 3. Aliphatic C—H Groups: 5–0 ppm

The remainder of the spectrum, from approximately 5 to 0 ppm, is where aliphatic C—H groups have their resonances. We further see that peaks for aliphatic fragments having one adjacent heteroatom or electron-withdrawing group are in the downfield portion of the region (2–5 ppm), whereas saturated C—H resonances appear farther upfield, between 0 and 2 ppm. When two electron-withdrawing groups are attached to a CH or $CH_2$ fragment, the resonance may be in the "alkene" region, between 5 and 6 ppm. A detailed summary for resonances in the aliphatic region (5–0 ppm) is presented in Figure 3.21 and Table 3.3.

**Table 3.3**

Approximate chemical shifts for protons bonded to aliphatic carbon atoms

| Methyl protons | | Methylene protons | | Methine protons | |
|---|---|---|---|---|---|
| Proton | $\delta$, ppm | Proton | $\delta$, ppm | Proton | $\delta$, ppm |
| $CH_3-C$ | 0.9 | $-CH_2-C$ | 1.3 | $-CH-C$ | 1.5 |
| $CH_3-C-C=C$ | 1.1 | $-CH_2-C-C=C$ | 1.7 | | |
| $CH_3-C-O$ | 1.4 | $-CH_2-C-O$ | 1.9 | $-CH-C-O$ | 2.0 |
| $CH_3-C=C$ | 1.6 | $-CH_2-C=C$ | 2.3 | $-CH-C=C$ | 2.2 |
| $CH_3-Ar$ | 2.3 | $-CH_2-Ar$ | 2.7 | $-CH-Ar$ | 3.0 |
| $CH_3-CO-R$ | 2.2 | $-CH_2-CO-R$ | 2.4 | $-CH-CO-R$ | 2.7 |
| $CH_3-CO-Ar$ | 2.6 | $-CH_2-CO-Ar$ | 2.9 | $-CH-CO-Ar$ | 3.5 |
| $CH_3-CO-O-R$ | 2.0 | $-CH_2-CO-O-R$ | 2.2 | $-CH_2-CO-O-R$ | 2.5 |
| $CH_3-CO-O-Ar$ | 2.4 | | | | |
| $CH_3-O-R$ | 3.3 | $-CH_2-O-R$ | 3.4 | $-CH-O-R$ | 3.7 |
| $CH_3-O-H$ | 3.5 | $-CH_2-O-H$ | 3.6 | $-CH-O-H$ | 3.9 |
| $CH_3-OAr$ | 3.8 | $-CH_2-OAr$ | 4.3 | $-CH-OAr$ | 4.5 |
| $CH_3-O-CO-R$ | 3.7 | $-CH_2-O-CO-R$ | 4.1 | $-CH-O-CO-R$ | 4.8 |
| $CH_3-N$ | 2.3 | $-CH_2-N$ | 2.5 | $-CH-N$ | 2.8 |
| $CH_3-NO_2$ | | $-CH_2-NO_2$ | 4.4 | $-CH-NO_2$ | 4.7 |
| $CH_3-C-NO_2$ | 1.6 | | | | |
| $CH_3-C=C-CO$ | 2.0 | | | | |
| $CH_3-C-Cl$ | 1.4 | $-CH_2-C-Cl$ | 1.8 | $-CH-C-Cl$ | 2.0 |
| $CH_3-C-Br$ | 1.8 | $-CH_2-C-Br$ | 1.8 | $-CH-C-Br$ | 1.9 |
| $CH_3-Cl$ | 3.0 | $-CH_2-Cl$ | 3.4 | $-CH-Cl$ | 4.0 |
| $CH_3-Br$ | 2.7 | $-CH_2-Br$ | 3.4 | $-CH-Br$ | 4.1 |

It is often simple to assign a feature in the region 0–5 ppm if the resonance does not overlap with any other, because the theoretical chemical-shift range for each type of C—H fragment is narrow (Figure 3.21); and only rarely will a group absorb outside the indicated range. Furthermore, for assigning peaks in the region from 2 to 5 ppm, data from the IR and mass spectra will narrow the possibilities. This is why you should look for heteroatoms and functional groups before attempting the interpretation of the NMR spectrum (see Figure 1.3). If the IR spectrum shows the presence of a carbonyl group, for example, then assignment of a resonance at 2.3 ppm as a $CH_2-C=O$ group would be very reasonable.

The last statement illustrates another point, too. The assignment of isolated resonances in the region 0 to 5 ppm is often straightforward if you consider the integrated intensity for the feature. Thus a resonance at 2.3 ppm, along with the appropriate IR data, suggests the presence of either a $CH_3-C=O$ or a $CH_2-C=O$ group. If the integrated intensity of the resonance were 3, you could be fairly certain that the fragment contains a methyl group. If the relative intensity were 6, however, then the fragment might contain *two* methyl or *three* methylene groups. Remember, in order for all such groups to have the same chemical shift, they must all be in the same chemical environments. For three groups to be chemically equivalent in a molecule with a low molecular weight ( < 200 excluding halogens), the compound will probably have a high symmetry, hence a simple NMR spectrum. You can see that if you assign a methine group to a resonance with a high relative-integration ratio, there will have to be an inordinate number of such groups, all in chemically equivalent positions. Such a situation is so unlikely that you can often eliminate certain groups from further consideration.

### 4. Potential Problems

In assigning the different chemical shifts, you need to be alert to certain pitfalls. First, several kinds of protons can appear over a wide range in the spectrum. They are included in the top portion of Figure 3.6 and comprise molecules having protons bound to N or O. Fortunately, most of the protons in that category exchange with $D_2O$; so you can do a simple experiment to confirm your assignment.

A second problem is that you cannot always tell whether a peak is actually from your compound, or arises instead from solvent that is not completely deuteriated. This is more of a problem when you are working with extremely dilute solutions on a Fourier-transform instrument, where the proton impurities from solvent may be at a concentration similar to that for the compound under investigation. Experimentally, you can either raise the concentration of the sample or change solvents, in order to distinguish between resonances for the compound and those of solvent.

Finally, when there are many similar aliphatic fragments in a molecule, the peaks in the region between 0 and 2 ppm are often not interpretable; and either the $^{13}C$ NMR spectrum or the mass spectrum may be more valuable for discovering the structure of the compound.

## C. Spin-spin Splitting

Following the assignment of the resonances to appropriate fragments based on the chemical shifts and integrated intensities, you need to begin to construct the carbon skeleton by using the spin-spin splitting to deduce which groups are adjacent to each other. From the multiplicity of each feature, you first calculate the number of neighboring protons, $N$, since you know that $N = M - 1$ (for a doublet, $M = 2$; for a triplet, $M = 3$; etc.). You then look for a feature having $N$ protons and the same $J$ value. Those two groups are adjacent. Continue the process until you have accounted for all of the NMR signals, and you will have generated the structure.

The advantage to having first assigned the type of proton to each set of resonances is that you know what pieces you have to work with as you put them together to generate the molecular structure. The best way to illustrate the procedure is by example, and several are presented in Section IV. Before we look at the examples, there are three general points to consider about spin-spin splitting patterns.

(1) *There are only a few splitting patterns for certain C—H fragments.* For example, a quintet at 2.3 ppm cannot be assigned as a $CH_2$—C=O group, because no more than three protons can be adjacent to a methylene group in this environment (i.e. $CH_3$—$CH_2$—C=O), which would make the feature at 2.3 ppm appear as a quartet ($3 + 1 = 4$). Thus, as you consider the splitting patterns, do not hesitate to reevaluate chemical-shift assignments if necessary.

(2) *In the aromatic region, significant splitting occurs only among the resonances in that portion of the spectrum*, because any protons on a side chain are too far removed to undergo coupling with the ring protons (Figure 3.22).

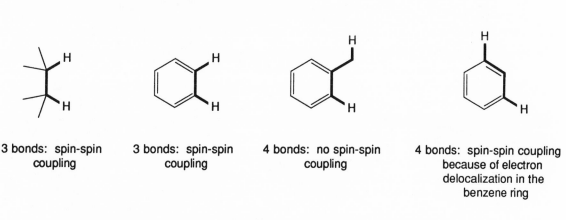

3 bonds: spin-spin coupling

3 bonds: spin-spin coupling

4 bonds: no spin-spin coupling

4 bonds: spin-spin coupling because of electron delocalization in the benzene ring

**Figure 3.22**
Effect of intervening bonds on the observation of spin-spin coupling.

If a compound has an aromatic ring, examine the splitting pattern of the resonances to confirm or identify the substitution pattern of the ring. Although the splitting among protons on the ring are often not first-order, sometimes the symmetry of the substitution pattern generates pseudo-first-order spectra that are easily recognized. For example, the proton resonances of para-disubstituted phenyl rings on which the substituents are electronically very different appear as two doublets (Figure 3.23, page 82) because to a first approximation, the protons resemble a —CH—CH— system (*cf.* Figure 3.18a).

Highly substituted phenyl rings are easily treated as first-order spectra, because there are few protons and the individual coupling constants are readily measured. For

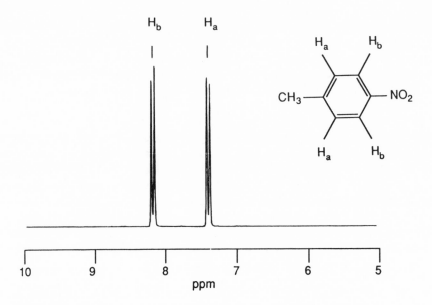

**Figure 3.23**
Proton NMR spectrum of *p*-nitrotoluene in the aromatic region, showing the first-order behavior of the splitting.

example, the spectrum shown in Figure 3.24 is assigned to a 1,2,4-trisubstituted benzene compound.

Mono- or disubstituted benzene rings can present either very simple or very complicated spectra. Complicated spectra are obtained when the proton environments are rendered very different because of the substituents; some examples are shown in Figure 3.25. Very simple spectra are observed when the chemical shifts for the different protons happen to be nearly the same; in such spectra, peaks in the aromatic region can appear as a singlet (Figure 3.26, page 84) or as a very symmetric pattern. Resonances for *ortho*-disubstituted rings can even (though rarely) appear as two doublets, as in *para*-disubstituted compounds.

In summary, the appearance of the aromatic region of an NMR spectrum depends only on the spin-spin splitting caused by the other protons on the aromatic ring. Since the chemical-shift differences are small, many of the patterns you see will not be first-order and can therefore be uninterpretable. Use the integrated intensity of the peaks in the aromatic region to clarify any ambiguity. For example, a peak or group of peaks in the aromatic region with a relative intensity of "5" usually indicates that the molecule has a mono-substituted phenyl ring, whatever the appearance of the NMR signals may be.

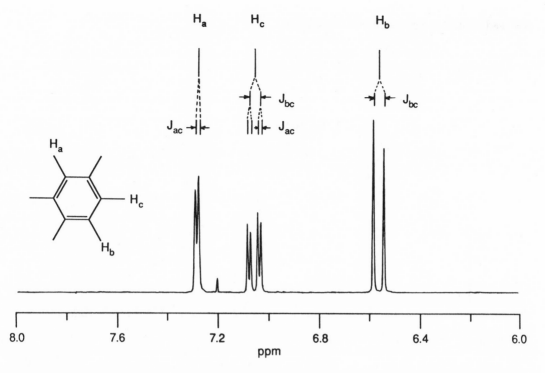

**Figure 3.24**
NMR spectrum of a typical 1,2,4-trisubstituted benzene compound showing assignments and splittings.

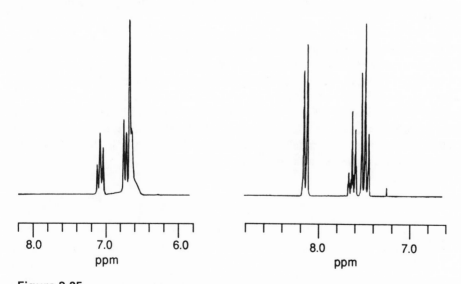

**Figure 3.25**
Proton NMR spectra in the aromatic region for *m*-cresol (left) and benzoic acid (right), illustrating the complicated splitting patterns that may be observed.

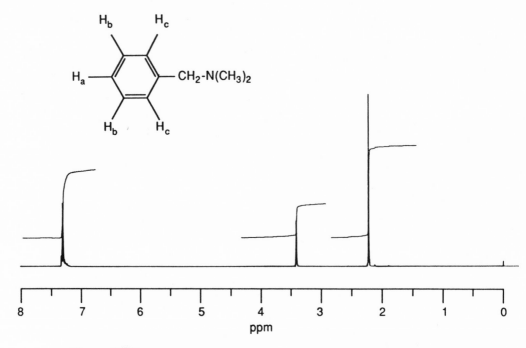

**Figure 3.26**
Proton NMR spectrum of dimethylbenzylamine. Note that all of the aromatic protons appear at the same chemical-shift values, although there are three types of protons.

(3) *Spin-spin splitting patterns for an alkene can be very complicated, because there are sometimes several different coupling constants to consider.* Not only do the alkene protons split each other, but they couple to attached alkyl groups. The latter coupling is almost always first-order, because the chemical shifts for the alkenyl and alkyl protons are so different. It is among the protons on the double bond that second-order effects are observed, and the second-order effects can obscure the coupling from attached alkyl groups. Fortunately, the substitution pattern of a double bond is often easily deduced from the IR spectrum.

## IV. EXAMPLES

Because of the importance of NMR spectroscopy in determining the structure of organic molecules, it is instructive to look at the interpretation process as applied to specific spectra.

### Compound 1

We will begin with the spectrum shown in Figure 3.27. The IR spectrum shows that this compound has an ester carbonyl group, and there are no heteroatoms other than oxygen.

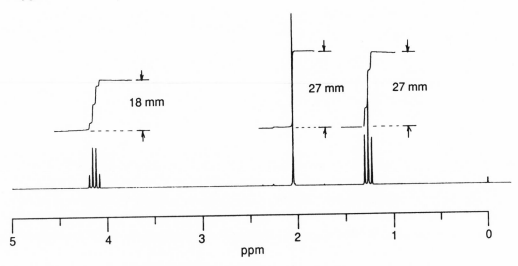

**Figure 3.27**
Proton NMR spectrum for Compound 1.

From the heights of the integration curves above each of the three features, we find the proton *ratio* to be 2:3:3. If we had the mass spectrum and knew the molecular formula, we would be able to calculate the actual number of protons for each feature.

Using the chemical-shift values, we can begin making assignments. Looking at Table 3.3, we see that the feature farthest downfield is most likely an aliphatic fragment bonded to an oxygen atom. We can rule out other possibilities, since we know that this compound contains only C, H, and O atoms. We are left to decide whether the group is a methyl, methylene, or methine. We can rule out a methyl group, because a $CH_3-O-$ group would appear as a singlet, and the feature at 4.13 ppm is clearly a quartet. We will have to come back to decide between the choices of methylene or methine when we have looked at the rest of the spectrum. The next feature is the one at 2.07 ppm. Using Table 3.3, we can assign this resonance to an aliphatic fragment adjacent to a carbonyl group. The fact that the $\delta$ 2.07 resonance is a singlet with a relative intensity of "3" strongly suggests that it results from a methyl group. Finally, the feature farthest upfield, at 1.28 ppm, is an aliphatic hydrocarbon group. Assuming our assignment of the singlet at 2.07 ppm as a $CH_3-C=O$ group is correct, we can assign the triplet at 1.28 ppm as a methyl group too, because the number of protons for that triplet is the same as the number of protons in the feature absorbing at 2.07 ppm. We can now be more specific about the quartet at 4.13 ppm, assigning it as a methylene fragment, since the integrated intensity of the quartet is "2." You can summarize the data by creating a table like Table 3.4.

**Table 3.4**
Summary of structural data for Compound 1

| Chemical shift ($\delta$) | Integrated intensity | Assignment | Multiplicity | $J$ (Hz) | # adjacent protons ($N$) |
|---|---|---|---|---|---|
| 4.13 | 2 | $CH_2-O$ | quartet | 7 | 3 |
| 2.07 | 3 | $CH_3-C=O$ | singlet | — | 0 |
| 1.28 | 3 | $CH_3-$ | triplet | 7 | 2 |

To assemble the molecule from the fragments in the table, we now make use of the spin-spin splitting patterns. This example can be solved by inspection, because there are only two resonances that show splitting; so the corresponding fragments must be adjacent. Thus, we generate the fragment $CH_3-CH_2-O$, which can be connected to $CH_3-C=O$, since we also know (from the IR spectrum) that the compound has an ester functionality. The structure is therefore

$$CH_3-CH_2-O-\overset{\overset{\textstyle O}{\|}}{C}-CH_3 \qquad \text{ethyl acetate} \qquad (3.14)$$

### Compound 2

The NMR spectrum for this compound is shown in Figure 3.28. The IR spectrum indicates that the molecule is an acid chloride, and the mass spectrum confirms the presence of the chlorine atom. Furthermore, the mass spectrum indicates a molecular formula with seven protons.

You can see immediately that the intensity ratio of $2:2:3$ calculated from the integration curve corresponds to the absolute numbers of protons for each feature ($2 + 2 + 3 = 7$).

Looking at the chemical shifts, you would assign the feature at 2.85 ppm to a methylene group (two protons) adjacent to an electron-withdrawing group. Using the data from the IR spectrum, you know that the electron-withdrawing group is the carbonyl moiety of an acid chloride. Using Table 3.3, you might be tempted to think that the triplet is too far downfield for that assignment to be correct, but you have to remember that the presence of the chlorine atom will tend to deshield the methylene protons more than a simple ketone carbonyl group would; so the assignment is reasonable. You can now assign the feature at 1.76 ppm to an aliphatic methylene

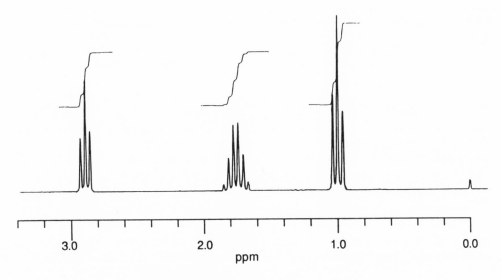

**Figure 3.28**
Proton NMR spectrum for Compound 2.

group (two protons), and the triplet at 1.0 ppm to an aliphatic methyl group (three protons), since both features fall in the aliphatic hydrocarbon region. Summarize the data as in Table 3.5.

**Table 3.5**
Summary of structural data for Compound 2

| Chemical shift ($\delta$) | Integrated intensity | Assignment | Multiplicity | $J$ (Hz) | # adjacent protons ($N$) |
|---|---|---|---|---|---|
| 2.85 | 2 | $CH_2$—C=O | triplet | 7 | 2 |
| 1.76 | 2 | $CH_2$ | sextet | 7 | 5 |
| 1.0 | 3 | $CH_3$ | triplet | 7 | 2 |

Again, we can solve this problem easily. If the methylene group absorbing at 2.85 ppm were adjacent to the methyl group absorbing at 1.0 ppm, we would not be able to fit the other methylene group into the formulation. Obviously, the methylene group absorbing at 1.76 ppm must be between the other two groups. Furthermore, the fact that the $\delta$ 1.76 feature is split into what appears to be a sextet says that there must be five protons on the adjacent carbon atoms (5 + 1 = 6). The formulation must therefore be:

$$CH_3—CH_2—CH_2—\overset{\overset{\displaystyle O}{\displaystyle \|}}{C}—Cl \qquad \text{butanoyl chloride} \qquad (3.15)$$

Before we leave Compound 2, look more closely at the "sextet" appearing at 1.76 ppm (Figure 3.29). If we consider the splitting of the resonance for the methylene group as a two-step process, we see that the protons on the adjacent methylene group first split the resonance into a triplet (2 + 1 = 3). Then the protons on the methyl group split each of those peaks into quartets. Here the two coupling constants are so close as to be nearly indistinguishable. However, if you look closely, you can see that the peaks are not exactly symmetric. Apparently identical coupling constants will be encountered for many compounds that contain an aliphatic chain, because the C—H environment is so similar. Thus, the $N + 1$ rule will hold, where $N$ is the *total* number of protons on the adjacent carbon atoms, even though all the protons are *not* chemically or magnetically equivalent.

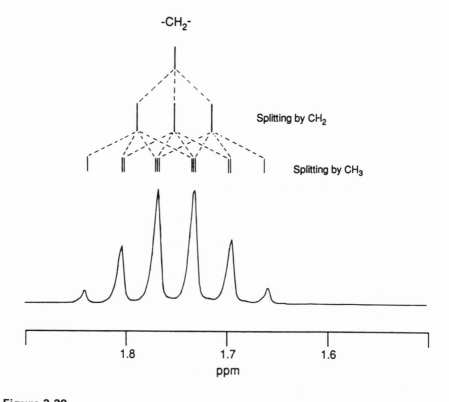

**Figure 3.29**
Expanded proton NMR spectrum for the sextet feature for Compound 2, showing the two slightly different coupling constants.

### Compound 3

The spectrum for Compound 3 is shown in Figure 3.30. The infrared spectrum shows an ester carbonyl group conjugated to a double bond. Let us assume, however, that we have not been able to deduce the configuration of the double bond.

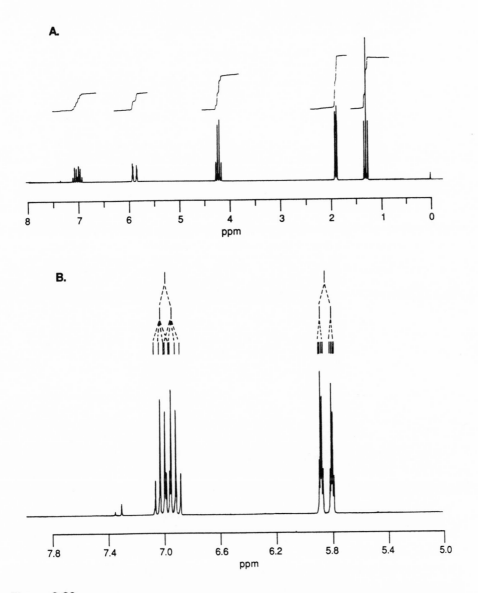

**Figure 3.30**
Proton NMR spectrum for Compound 3. A, normal range spectrum; B, expanded spectrum to show spin-spin splitting of the alkenyl protons.

The integrated intensities for the features are in the ratio of $1:1:2:3:3$. We do not yet know if these values correspond to absolute numbers of protons, because we do not have the mass spectrum.

Since we know that the compound contains an alkene group, we can assign the features at 6.95 and 5.85 to hydrogen atoms attached to the double bond. Since an $\alpha,\beta$-unsaturated ester can have at most three protons on the alkene portion of the molecule, you can assume, reasonably, that the relative number of protons in the alkene region $(1 + 1 = 2)$ corresponds also to the absolute number. You can now assign the feature at 4.2 ppm to a methylene group (two protons) adjacent to an oxygen atom (Table 3.3), the feature at 1.9 ppm as a methyl group (three protons) attached to the alkene (Table 3.3), and the triplet at 1.3 ppm as an aliphatic methyl group (three protons).

To assign the coupling constants, you have to study the spectrum carefully. The easiest $J$-values to calculate are those for the quartet at 4.2 ppm and the triplet at 1.3 ppm. The feature at 1.9 ppm appears to be two doublets, and the feature at 5.85 ppm appears to consist of two quartets. The most difficult assignment is that of the proton resonance at 6.95 ppm. It also appears to be two quartets, as indicated by the lines drawn above the expanded region in Figure 3.30. We can summarize the data as in Table 3.6.

**Table 3.6**
Summary of structural data for Compound 3

| Chemical shift ($\delta$) | Integrated intensity | Assignment | Multiplicity | $J$ (Hz) | #adjacent protons ($N$) |
|---|---|---|---|---|---|
| 6.95 | 1 | C=C—H | two quartets | 6,15 | 1,3 |
| 5.85 | 1 | C=C—H | two quartets | 2,15 | 1,3 |
| 4.2 | 2 | $CH_2$—O | quartet | 7 | 3 |
| 1.9 | 3 | $CH_3$—C=C | two doublets | 2,6 | 1,1 |
| 1.3 | 3 | $CH_3$ | triplet | 7 | 2 |

It is actually not difficult to assemble the structure from the collected data. Clearly the most difficult part is to meld the splitting patterns. The easiest connection is the methyl group ($\delta = 1.3$ ppm) with the $CH_2$—O fragment, because the splitting patterns are as expected and the $J$ values are identical. We know from the IR spectrum that this compound is an $\alpha,\beta$-unsaturated ester; so we already have the fragment

$$\text{C}=\text{C}-\overset{\overset{\textstyle O}{\textstyle \|}}{\text{C}}-\text{O}-\text{CH}_2\text{-}\text{CH}_3 \tag{3.16}$$

We know from the chemical-shift data that there is a methyl group attached to the alkene, but the question is where. We can draw three possible configurations, as in Figure 3.31, and predict the expected $J$ values by using the data in Figure 3.16.

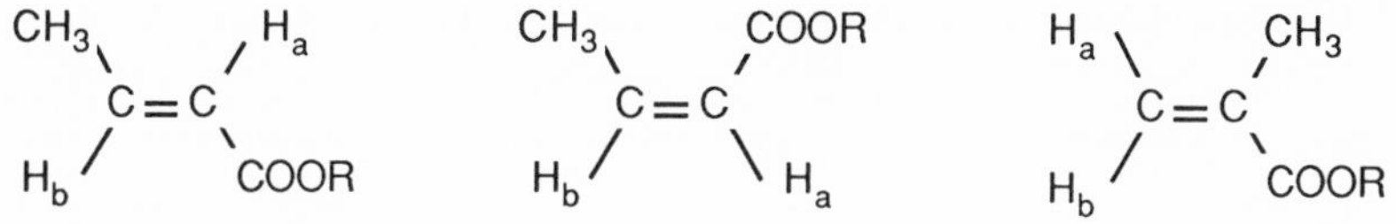

$$CH_3 \diagdown \atop H_b \diagup C = C \diagup H_a \atop \diagdown COOR \qquad\qquad CH_3 \diagdown \atop H_b \diagup C = C \diagup COOR \atop \diagdown H_a \qquad\qquad H_a \diagdown \atop H_b \diagup C = C \diagup CH_3 \atop \diagdown COOR$$

| $J_{CH_3-H_a}$ | 4-10 | 4-10 | 0-3 |
|---|---|---|---|
| $J_{CH_3-H_b}$ | 0-3 | 0-3 | 0-3 |
| $J_{H_a-H_b}$ | 12-18 | 6-12 | 0-3 |

**Figure 3.31**
Three possible substructures in Compound 3. All values are in Hz.

Clearly the values predicted for the *trans* configuration match most closely the data obtained from the spectrum; so we can now draw the complete structure for the compound:

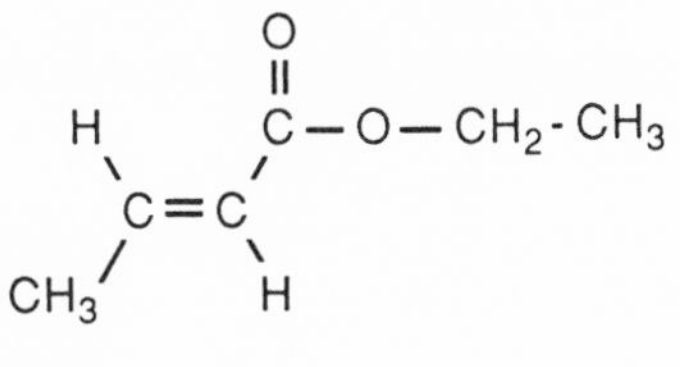

$$H \diagdown \atop CH_3 \diagup C = C \diagup {\overset{\displaystyle O \atop \|}{C}} - O - CH_2 \cdot CH_3 \atop \diagdown H \qquad \text{ethyl crotonate} \qquad (3.17)$$

### Compound 4

The proton NMR spectrum for this compound is illustrated in Figure 3.32 (on page 92). The IR spectrum shows that the molecule contains a carboxylic acid group, conjugated to either an aromatic ring or alkene.

Starting with the integrated intensities, we measure the ratios 1:1:1:2:3 from left to right.

Assigning the chemical shifts, we begin with the resonance farthest downfield. This peak is easily assigned as the carboxylic-acid proton, because of the position and broadness of the resonance signal. It is better seen in the expanded spectrum in the lower part of Figure 3.32. If we assume there is only one such group in the molecule, then we know that the integration ratios also correspond to the absolute number of protons for each feature.

The peaks in the region from 7.0 to 8.2 ppm correspond most likely to phenyl protons, judging from their chemical shifts. We know from the IR spectrum that there

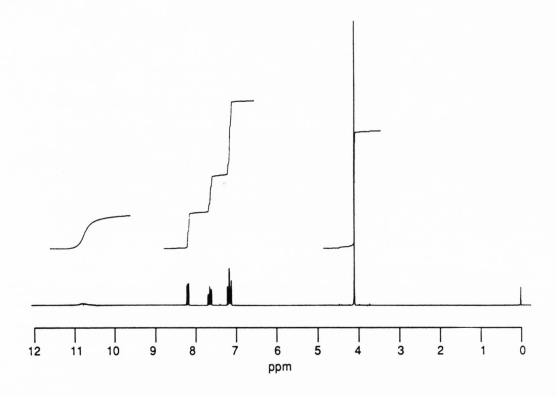

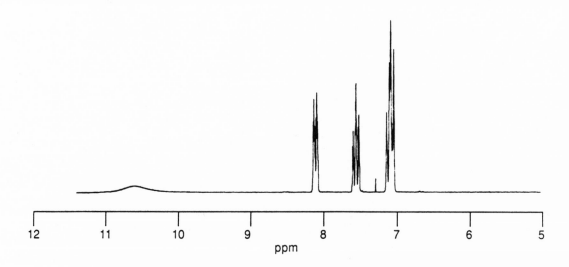

**Figure 3.32**
Proton NMR spectrum for Compound 4. Top, normal range spectrum; bottom, expanded spectrum of the aromatic proton region.

is either an aromatic ring *or* a double bond, but resonances for the latter functional group usually appear farther upfield (*cf.* Figure 3.30). If we assign the resonances between 7.0 and 8.2 ppm to phenyl protons, then we should consider the integration curve for the whole region. Thus, the total number of protons on the ring is four, and we can be certain that the benzene ring is disubstituted.

Finally, we can attribute the resonance at 4.1 ppm to a methyl group, which from its chemical shift (Figure 3.6) must be attached to an electron-withdrawing group. The electron-withdrawing group cannot be the phenyl ring itself, since $CH_3$—Ar absorptions usually appear around 2.3 ppm. More likely, given the data in Figure 3.21, the methyl group is attached to an oxygen atom that is bonded to the aromatic ring.

The only spin-spin splitting observed for this compound is in the aromatic region among the protons bonded to the phenyl ring. We can summarize the data as shown in Table 3.7.

**Table 3.7**
Summary of structural data for Compound 4

| Chemical shift ($\delta$) | Integrated intensity | Assignment | Multiplicity | $J$ (Hz) | #adjacent protons ($N$) |
|---|---|---|---|---|---|
| 10.7 | 1 | COOH | singlet | — | 0 |
| 7–8.2 | 4 | Ar—H | multiplet | — | — |
| 4.1 | 3 | $CH_3$—O | singlet | — | 0 |

We can be fairly certain from the lack of symmetry among the peaks in the aromatic region that the substituents (COOH and $OCH_3$) are not *para* to each other (*cf.* Figure 3.24); so there are only two possibilities:

$$(3.18)$$

Because the splitting patterns of the ring protons are not first-order, we will not attempt a detailed analysis using the $J$ values. Instead, we will consider a more qualitative approach, by asking what we would expect the splitting pattern to look like for the proton *ortho* to the carboxy group in each structure (proton *a* in 3.18). For *ortho*-anisic acid, $H_a$ "sees" the effects of other protons that are *o*, *m*, and *p*. The attendant $J$ values (Figure 3.16) are 6–10, 1–3, and 0–1 Hz, respectively. For *meta*-anisic acid, $H_a$ is influenced by protons that are only *m* and *p*, since the *o*-positions are substituted. Therefore, the $H_a$ resonance will be split only very slightly (1–3 and

0–1 Hz), and should appear more like a singlet than a complex multiplet. Since the actual observed features in the aromatic region are all quite complex, we conclude that the structure of Compound 4 is most likely *ortho*-anisic acid:

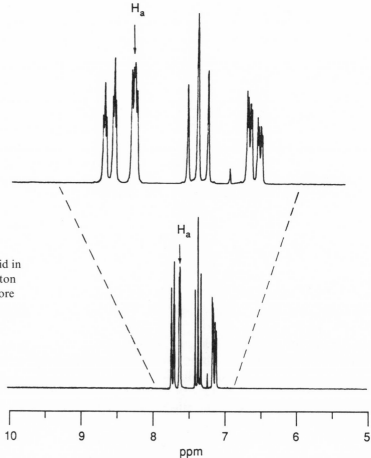

o-anisic acid  (3.19)

To confirm that our qualitative analysis is reasonable, the NMR spectrum of the aromatic region for *meta*-anisic acid is shown in Figure 3.33. The resonance for $H_a$, although not quite a singlet, is split much less than the other signals.

**Figure 3.33**
Proton NMR spectrum of *m*-anisic acid in the aromatic region, showing that proton $H_a$ gives rise to a peak that appears more like a singlet than those peaks for the other aromatic protons.

## References

The following texts either give additional correlation charts and tables, or more detailed discussion about topics covered in this chapter.

R. J. Abraham and P. Loftus, *Proton and $^{13}C$ Spectroscopy: An Integrated Approach*. London: Heyden, 1978.

T. C. Farrar and E. D. Becker, *Pulse and Fourier-Transform NMR*. New York: Academic Press, 1971.

V. M. Parikh, *Absorption Spectroscopy of Organic Molecules*. Reading, MA: Addison-Wesley, 1974.

D. L. Pavia, G. M. Lampman, and G. S. Kriz, Jr., *Introduction to Spectroscopy*. Philadelphia: Saunders, 1977.

E. Pretsch, T. Clerc, J. Seibl, and W. Simon, *Tables of Spectral Data for Structure Determination of Organic Compounds*. Berlin: Springer-Verlag, 1983.

R. M. Silverstein, G. C. Bassler, and T. C. Morrill, *Spectrometric Identification of Organic Compounds*. New York: Wiley, 4th ed., 1978.

# CHAPTER 4

# ¹³C Nuclear Magnetic Resonance Spectroscopy

Carbon is the ubiquitous building block of organic molecules; so analytical techniques that can directly probe the carbon-atom environments within a molecule are especially useful. Nuclear magnetic resonance spectra of carbon nuclei were observed only a very few years after the advent of proton NMR spectroscopy; however, not until after development of computerized data-collection systems did ¹³C NMR spectroscopy became a routine technique in organic chemistry. Now, after three decades, it is apparent that we are really only beginning to develop the capabilities of this tool.

## I. THEORY

An easy way to approach ¹³C NMR spectroscopy is to highlight the similarities and differences between ¹³C and ¹H NMR spectra.

Starting with the similarities, the absorption process that gives rise to the NMR signals is identical. As described in Chapter 3, atomic nuclei containing an odd number of protons and neutrons display a nuclear magnetic moment. ¹²C has no unpaired nuclear spins; however, ¹³C has an extra neutron, for which there is an associated spin-change transition when the nucleus is placed in a magnetic field. The frequency required to cause resonance is given by:

$$v = \frac{\gamma H_0}{2\pi}.$$  (4.1)

Another similarity is that, as is true for proton resonances, individual carbon-atom resonances are slightly different from one another, and lead to a range of chemical shifts (see Section 3.I). The chemical shifts arise because of the electron distribution within the carbon framework; and electronegativity, hybridization, and anisotropy all play a role in influencing the frequency at which resonance occurs. $^{13}$C chemical shifts in most organic molecules cover a range of 250 ppm, and are compared to the resonances for the methyl carbon atoms of tetramethylsilane, TMS, $(CH_3)_4Si$. A compilation of $^{13}$C chemical shifts is presented in Figure 4.1.

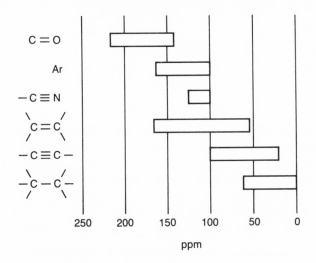

**Figure 4.1**
General chemical-shift ranges for carbon atoms.

Now let us consider the *differences* between proton and carbon NMR spectroscopy. The first difference is that the magnetogyric ratio, $\gamma$, for carbon is about a fourth of that for the proton; so the frequency needed to cause resonance (Equation 4.1), given the same field strength as used for a proton NMR spectrum, will be quite different. As a result of this difference, the proton and carbon resonances do not appear together in the same region of the radiofrequency spectrum.

Another contrast between $^1$H and $^{13}$C NMR spectra is in the strength of the signals observed for the two. For proton NMR, nearly every hydrogen atom contributes to the resonance signal, since the isotopic abundance of $^1$H is almost 100 percent. For carbon, the natural abundance of $^{13}$C is only 1.1 percent, which means that most of the carbon nuclei are *not* involved in the resonance process. In addition, the sensitivity of a nucleus to the spin-change transition is proportional to $\gamma^3$; so even if the isotopic abundance of $^{13}$C were the same as that for $^1$H, the sensitivity for carbon would still be only 1/64 that for hydrogen. The lower magnetic sensitivity, coupled with the low abundance of $^{13}C$, means that the overall sensitivity for carbon is *6000 times* less than that for hydrogen.

The low isotopic abundance of $^{13}C$ manifests itself in another way. Recall from Chapter 3 that the presence of the neighboring protons leads to spin-spin coupling of the proton nuclei, which in turn can be used to deduce what fragments are adjacent to one another. For carbon, even if a particular atom is the $^{13}C$ isotope, the chances that any neighboring carbon atom is also $^{13}C$ is very small. Thus, $^{13}C$-$^{13}C$ splitting is not observed unless you synthesize a compound using starting materials that are isotopically enriched with $^{13}C$.

Another difference between carbon and proton NMR spectra is that the integrated intensities for carbon resonances are not accurate, because the relaxation times, $T_1$ and $T_2$, differ so much from carbon to carbon. Recall from Section 3.I that one pathway for a nucleus in the excited state (spin $= -\frac{1}{2}$) to relax to the ground state (spin $= +\frac{1}{2}$) is by spin-lattice relaxation. For carbon, the dominant mechanism for returning the nuclear spins to the ground state is by $^{13}C$-$^1H$ dipole-dipole relaxation. Since carbon atoms without attached protons do not have this pathway available, $T_1$ is lengthened, and the nucleus is easily "saturated"; that is, nuclei do not have time to relax before the next pulse of radiation. Thus, only a small signal is observed for those carbon nuclei. As you will see later, carbon atoms having no protons attached often give signals with very low intensities. The Nuclear Overhauser Effect (NOE), which we will discuss, also affects the intensities of the signals. *The result is that $^{13}C$ NMR spectra are not often integrated.*

The greatest difference between proton and carbon spectra is in their overall appearance. $^{13}C$ NMR spectra are usually presented as a series of singlets, which are unusual in proton spectra, except for spectra of the simplest compounds. In order to understand why, we have to consider the effects of $^{13}C$-$^1H$ coupling.

The low abundance of $^{13}C$ accounts for the absence of $^{13}C$-$^{13}C$ spin-spin splitting. The same is true for the lack of $^{13}C$-$^1H$ splitting in the proton spectrum: the likelihood that the proton is bound (or near) a $^{13}C$ atom is quite low. However, the converse is not true. If there is a $^{13}C$ atom in a molecule, there will almost certainly be protons within three or four bonds that can undergo spin-spin coupling. In fact, there will probably be several different $J$-values, as summarized in Figure 4.2.

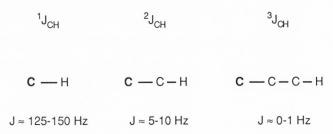

**Figure 4.2**
Effect of intervening bonds on the $^1H$-$^{13}C$ coupling constant, $J_{CH}$.

## A. The Proton-coupled ¹³C NMR Spectrum

To begin to see what a typical ¹³C NMR spectrum looks like under various conditions, let us consider the spectrum of allyl chloride, in Figure 4.3.

Allyl chloride has three different carbon atoms, each of which can couple to five protons. You would anticipate that there will be at least three different C—H coupling constants for each carbon atom, leading to complicated multiplets at each chemical-shift value. The splitting of each resonance into several peaks, in addition to the low sensitivity of the carbon nuclei, explains why the spectrum has small peaks

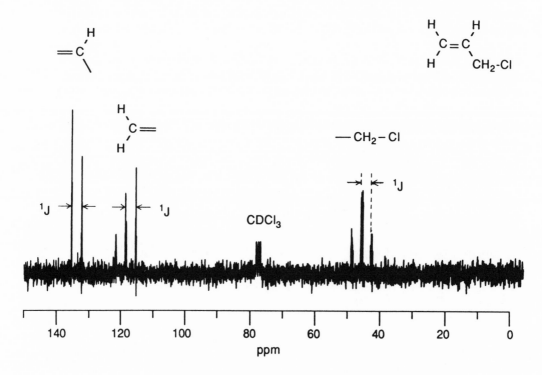

**Figure 4.3**
¹³C NMR spectrum of allyl chloride without decoupling.

relative to the baseline "noise" (although not all coupled $^{13}C$ NMR spectra are as noisy as this one). Fortunately, most $^{13}C$ spectra are not recorded with proton coupling, since otherwise spectra for compounds with even ten carbon atoms would be uninterpretable.

## B. The Broad-band Decoupled Spectrum

Suppose we now record the $^{13}C$ NMR spectrum of allyl chloride, except that in the process, we irradiate the sample with a second radiofrequency source covering the *proton* NMR range. This process is called "broad-band decoupling." In proton-decoupling experiments, the second irradiation saturates the protons and equalizes the populations of the two spin states. Thus, the spins of the protons are being rapidly converted from one energy state to the other, and the average spin state influencing the carbon resonances is constant. Recall from Chapter 3 that spin-spin coupling occurs because a nucleus experiences several different magnetic fields from the spin orientations of the neighboring nuclei (cf. Figure 3.14). If the spin orientations are averaged into one magnetic contribution, then no spin-spin splitting can occur. The influence of the decoupling of the protons on the $^{13}C$ spectrum of allyl chloride is shown in Figure 4.4 (on page 102).

In Figure 4.4, notice that the size or areas of the peaks in the spectrum appear greater than those for the peaks in Figure 4.3. That is, when the protons are irradiated during the decoupling process, you cannot explain the resulting appearance of the carbon resonances (Figure 4.4) simply by assuming that the peaks for the carbon atoms in the coupled spectrum (Figure 4.3) have been consolidated. The increase in peak size results because the Nuclear Overhauser Effect (NOE) is operational.

The NOE occurs whenever the population of the lower energy state ($s = \frac{1}{2}$) increases relative to that established by the Boltzmann distribution. For carbon nuclei, this increase usually occurs when the attached hydrogen nuclei are saturated by irradiation at their resonance frequencies. The carbon signals can theoretically increase three-fold in intensity, but not all carbon atoms are affected equally. The unequal increase is one reason that the integration of $^{13}C$ NMR signals is unreliable.

The broad-band decoupled spectrum is the type you will most often encounter if you use $^{13}C$ NMR spectra to analyze structure. Its simplicity is apparent, and the chemical shift for each carbon atom is easily identified. In fact, a resonance for every carbon atom can often be observed in a broad-band decoupled spectrum, even if there are 20 or 30 carbon atoms in the molecule. The NOE enhancement of each resonance also improves the signal-to-noise ratio substantially: the spectra in Figures 4.3 and 4.4

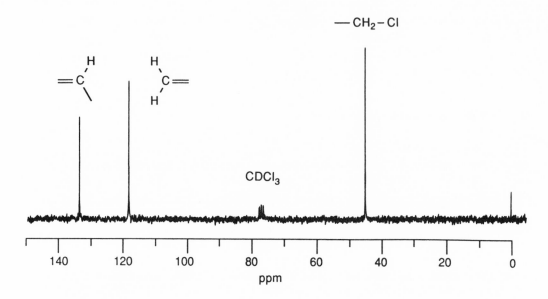

**Figure 4.4**
Broad-band decoupled ¹³C NMR spectrum of allyl chloride, showing the effect of complete proton decoupling.

were recorded under the same conditions, so that the increase in sensitivity for the decoupled spectrum is apparent.

The only disadvantage of the broad-band decoupled spectrum is that all coupling information has been lost, but this loss is a small price to pay, since coupling in the proton NMR spectrum is actually more useful for analyzing structures.

## C. The Off-resonance Decoupled Spectrum

Now let us consider a final experiment. This time, as with the broad-band decoupled spectrum, we irradiate the sample with a second radiofrequency source that covers the proton frequencies. Here, however, the principal frequency of the second source is 1000–2000 Hz *away* from the proton range; whereas for the broad-band decoupling, the principal frequency was in the middle of the proton range. Figure 4.5 shows what this type of spectrum, called an *off-resonance decoupled* spectrum, looks like.

The difference between a broad-band and an off-resonance decoupled spectrum is that there *is* spin-spin splitting in the latter. However, the observed *J*-value does not correspond to any real C—H coupling constant, but depends instead on how far away from the proton range the principal frequency is. The overall effect is to decouple

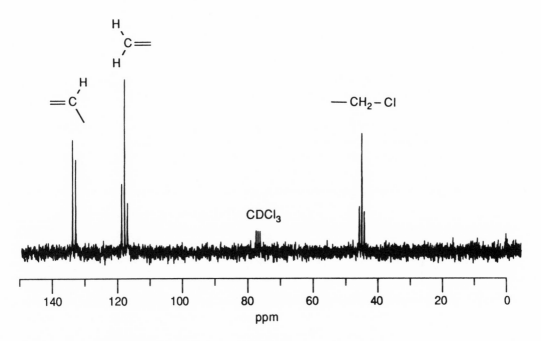

**Figure 4.5**
Off-resonance decoupled $^{13}C$ NMR spectrum of allyl chloride.

all but *the attached protons* from each carbon atom; so you observe *only* the one-bond coupling ($^1J_{CH}$, Figure 4.2). In this experiment, if the carbon resonance observed in the broad-band decoupled spectrum remains as a singlet, then there are no protons attached to that carbon atom. If the resonance appears as a doublet in the off-resonance decoupled spectrum, there is one proton attached; a triplet, two protons attached; and a quartet, three protons attached. You will never see more than four peaks (unless you are doing the spectrum of methane), because there will never be more than three hydrogen atoms attached to a carbon atom. Remember, in the off-resonance decoupled spectrum, all protons on neighboring carbon atoms have been decoupled, and so do not participate in spin-spin coupling.

The appearances of the different types of $^{13}C$ NMR spectra are summarized in Figure 4.6 (on page 104).

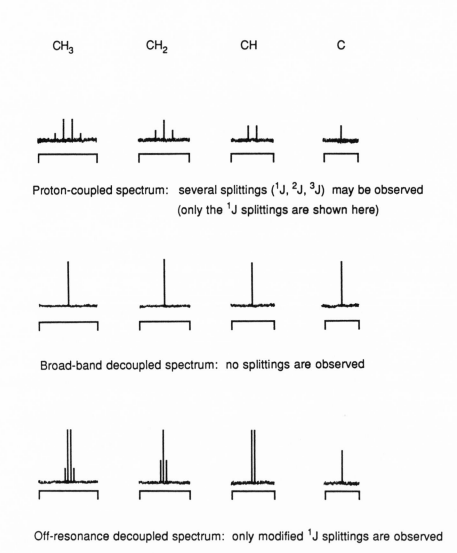

CH₃       CH₂       CH       C

Proton-coupled spectrum: several splittings ($^1$J, $^2$J, $^3$J) may be observed
(only the $^1$J splittings are shown here)

Broad-band decoupled spectrum: no splittings are observed

Off-resonance decoupled spectrum: only modified $^1$J splittings are observed

**Figure 4.6**
Appearance of $^{13}$C NMR signals under various decoupling conditions.

## II. EXPERIMENTAL CONSIDERATIONS

The NMR spectrometers used for running carbon spectra are often the same as those used for proton spectra, except that they must have computerized data-acquisition systems. The spectra are recorded using Fourier-transform techniques (summarized in Farrar's book listed at the end of this chapter), in which many scans of the frequency range are collected and averaged by the computer.

The spectrum is plotted over the range from approximately 250 to 0 ppm in the same manner as for proton spectra, with TMS on the right, and upfield and downfield relationships defined accordingly. Integrated intensities of the peaks are not usually calculated, for the reasons noted in the previous section.

The important data obtained from the $^{13}C$ NMR spectrum are the positions of the absorptions in ppm (the chemical shifts). In addition, you will want to record the off-resonance decoupled spectrum in order to find out how many hydrogen atoms are attached to each carbon atom.

To prepare a sample for a $^{13}C$ NMR spectrum, dissolve some of the compound in a solvent containing a small amount of the standard TMS, $(CH_3)_4Si$, and introduce the solution into the same type of thin-walled glass tube used to record the proton spectrum. The actual amount of compound needed for a $^{13}C$ spectrum is not critical, since you can simply perform more scans on the sample. However, using more compound ensures a better signal-to-noise resolution in the resulting spectrum.

Although solvents containing protons can be used for recording carbon NMR spectra, deuterochloroform $(CDCl_3)$ is probably the most widely used solvent; its single carbon atom is split into three peaks by coupling to the deuterium atom (deuterium has a nuclear spin I = 1). The use of a deuterated solvent allows good frequency control, because the spectrometer can "lock" onto the deuterium signal.

Because of the large excess of solvent relative to sample, and because the relative abundance of $^{13}C$ in the solvent is the same as in the sample, symmetric compounds should be used as solvents in order to minimize the number of extraneous peaks. Dioxane, dimethylsulfoxide (DMSO), benzene, and methanol have only one type of carbon, and so are used routinely. If the corresponding deuterated compound is used (for example, $C_6D_6$), then the carbon signal will appear as a triplet in the proton-decoupled spectrum because of the I = 1 spin of deuterium. Table 4.1 (on page 106) summarizes where $^{13}C$ solvent peaks are commonly observed.

**Table 4.1**

¹³C Chemical shifts for common solvents

| Solvent | | $\delta$ (ppm) |
|---------|------|----------------|
| Benzene | $C_6H_6$ | 128.5 |
| Carbon tetrachloride | $CCl_4$ | 96.0 |
| Cyclohexane | $C_6H_{12}$ | 26.9 |
| Deuterochloroform | $CDCl_3$ | 76.9 |
| Dimethylsulfoxide | $CH_3SOCH_3$ | 40.5 |
| Dioxane | $O(CH_2CH_2)_2O$ | 67.4 |
| Methanol | $CH_3OH$ | 49.3 |
| Methylene chloride | $CH_2Cl_2$ | 54.0 |

## III. INTERPRETATION OF THE ¹³C NMR SPECTRUM

There are two ways to approach interpreting a ¹³C NMR spectrum, depending on whether you are using it to identify functional groups, that is, in place of infrared spectroscopy, or using it to confirm previous assignments deduced from the IR or proton NMR spectra. We will look at the more general strategy first.

### A. Identification of Functional Groups

If you do not have the IR spectrum of the compound, then the ¹³C NMR spectrum provides a simple way to find out which carbon-containing functional groups are present. Because only the carbon atoms can be observed, the presence of functionalities such as O—H, S—H, N—H, $NH_2$, N=O, $NO_2$, S=O, and $SO_2$ cannot be detected directly. To interpret the spectrum, you first use the chemical-shift data, summarized in Figure 4.7, to find out what types of carbon atoms are present. You can then use the off-resonance decoupled spectrum to find the number of protons attached to each carbon atom.

#### 1. 250–160 ppm

As with proton NMR spectra, absorptions in the region farthest downfield from TMS indicate the presence of only a few compound types. In ¹³C spectra, the region from 160–250 ppm is where carbonyl carbon resonances appear. Remember that in the interpretation of IR spectra, the C=O functionality was the easiest to assign because of the position and intensity of its absorption band. Similarly for ¹³C NMR spectra, peaks in the region farthest downfield from TMS are readily assigned to specific

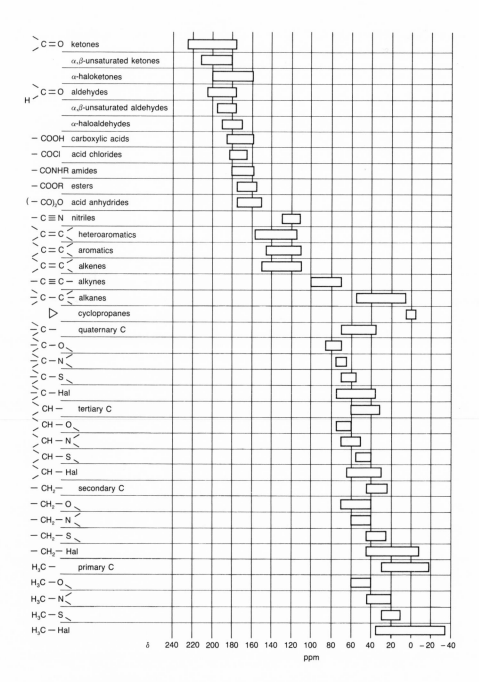

**Figure 4.7**
A detailed compilation of chemical shifts for $^{13}C$ NMR spectra.

carbonyl groups; however, as shown in Figure 4.8, the peak is often weak, because there are no protons attached (except for aldehydes); so the NOE enhancement is minimal.

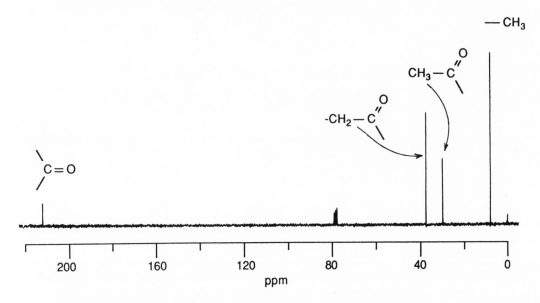

**Figure 4.8**
Broad-band decoupled ¹³C NMR spectrum of 2-butanone, showing the relatively low intensity for the carbonyl carbon resonance.

### 2.   160–100 ppm

The next region to examine is the one between 100–160 ppm, in which compounds having sp² and nitrile carbon atoms have resonances. Of course, the easiest way to distinguish between those two types of carbon atoms is to look for the unmistakable C≡N stretch in the IR spectrum. Also, a nitrile will give rise to only a single peak in the region from 100–160 ppm (Figure 4.9), which will be a singlet in the off-resonance decoupled spectrum.

Finding out whether the compound contains an aromatic ring or alkene functionality is sometimes not as simple. One way to find out whether the compound is an arene or alkene is to examine the out-of-plane C—H bending region in the IR spectrum. Another way is to use the proton NMR data, since often those two types of compounds are easily distinguished by means of their chemical shifts and integrated intensities.

Using only the ¹³C NMR data, you can tell if a terminal alkene is present, because one of the carbon resonances will appear as a triplet in the off-resonance decoupled spectrum, as shown in Figure 4.10. Remember, if there are *N* hydrogen atoms attached

**Figure 4.9**
Off-resonance decoupled $^{13}$C NMR spectrum of acetonitrile.

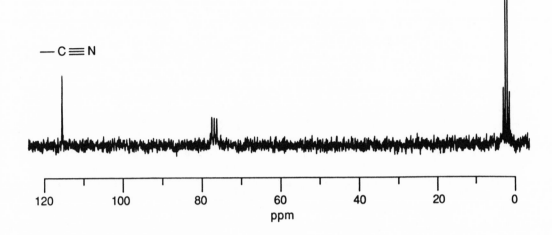

**Figure 4.10**
Off-resonance decoupled $^{13}$C NMR spectrum of 4-bromo-1-butene, showing the expected triplet at 118 ppm for the terminal $=CH_2$ group.

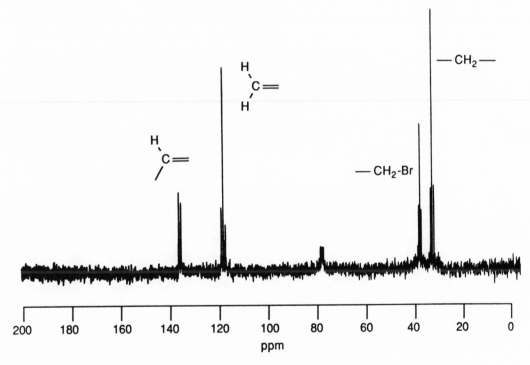

to the carbon atom, the off-resonance decoupled resonance will be split into $N + 1$ peaks. Since arenes can have at most one proton attached to each carbon atom, their resonances can appear only as singlets or doublets. Another way to distinguish arenes from alkenes is from the number of peaks. An unsymmetrically substituted benzene ring may have as many as six different carbon resonances, as illustrated in Figure 4.11. You have to be careful, however, because if the substitution pattern is symmetric, then only two or three peaks may be observed (Figure 4.12). Finally, alkenes usually exhibit an even number of peaks for the double-bonded carbon atoms, unless the molecule is symmetric or two of the carbon resonances fortuitously have the same chemical shift.

### 3. 100–70 ppm

The next region to examine is between 70 and 100 ppm. This region is where alkyne carbon-atom resonances appear, and the only overlap is with resonances for some types of 4° carbon atoms, attached to an oxygen atom. For each triple bond, there will be two carbon resonances, unless the compound is symmetric.

### 4. 70–0 ppm

The final region to study is between 0 and 70 ppm, in which $sp^3$ carbon-atom resonances appear. Assignment of the alkyl carbon resonances to specific fragments is sometimes difficult, since for each chemical-shift value there may be several possibilities. The off-resonance decoupled spectrum can be very helpful in narrowing the choices. For example, suppose you are studying a molecule containing no nitrogen or halogen atoms. Then a peak at 65 ppm could result from the presence of either an $RCH_2$—O or $R_2CH$—O group. If the peak at 65 ppm were split into a triplet in the off-resonance decoupled spectrum, then you would know that the $RCH_2$—O assignment is correct.

### 5. Potential Problems

Remember that as you examine the alkyl carbon resonances, you cannot discover the connectivity of the different C—H fragments, because all information about the adjacent groups is lost in the decoupling experiments. For example, the spectrum shown in Figure 4.13 (on page 112) has six signals, three of which are triplets in the off-resonance decoupled spectrum. You could propose either $CH_3$—$CH_2$—COO—$CH_2$—$CH_2$—$CH_3$ or $CH_3$—$CH_2$—$CH_2$—COO—$CH_2$—$CH_3$ as a reasonable structure. However, without using coupling information from the proton NMR spectrum, you would have a difficult time choosing the correct assignment.

Symmetry has been mentioned several times: you need to realize that only chemically nonequivalent carbon atoms will have different resonances. Unlike proton

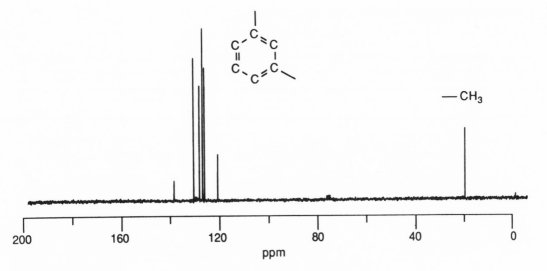

**Figure 4.11**
Broad-band decoupled $^{13}C$ NMR spectrum of *m*-bromotoluene, showing the nonequivalence of the aromatic-ring carbon atoms by the appearance of six peaks.

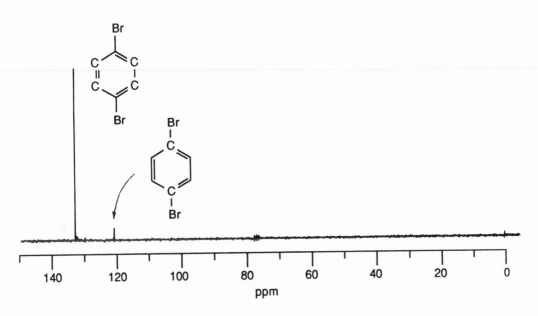

**Figure 4.12**
Broad-band decoupled $^{13}C$ NMR spectrum of a symmetric, disubstituted benzene compound, *p*-dibromobenzene.

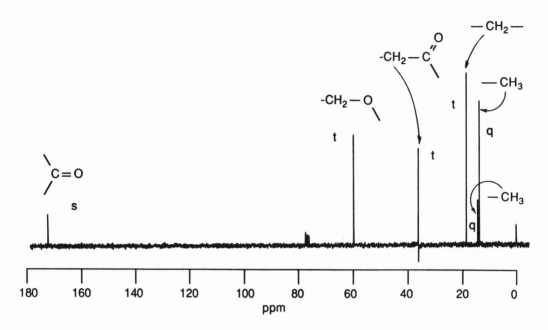

**Figure 4.13**

Broad-band decoupled ¹³C NMR spectrum of ethyl butyrate. The results of the off-resonance decoupled spectrum are summarized by the appropriate letter: s, singlet; t, triplet; q, quartet.

NMR, in which you have both chemical-shift data *and* integrated intensity values with which to work, you have only the chemical shifts for ¹³C NMR spectroscopy. Thus, if you had no other spectral data, and you were to look at the ¹³C spectrum shown in Figure 4.14, you could propose at least two structures consistent with the observed chemical shifts, namely, $(CH_3)_3C—OCH_3$ and $(CH_3)_2C(OCH_3)_2$. Note that there are only three types of carbon atoms in each structure. You would not have the same trouble with the proton NMR spectrum, however, because you could use the integrated intensities of the peaks to distinguish between the two possibilities.

In identifying either the functionality or the types of carbon fragments present in a molecule, you should always remember that you can use the IR or proton NMR spectra to check your assignments. Fortunately, you will rarely have just the ¹³C NMR spectrum.

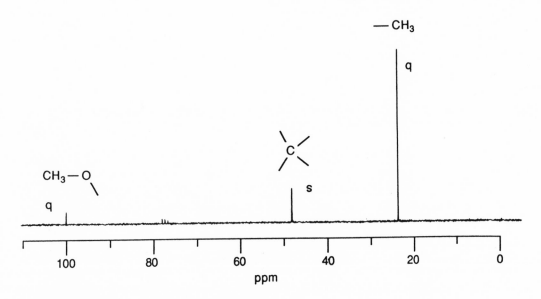

**Figure 4.14**
Broad-band decoupled ¹³C NMR spectrum of 2,2-dimethoxypropane. The results of the off-resonance decoupled spectrum are summarized by the appropriate letter: s, singlet; q, quartet.

## B. Confirmation of Previous Assignments

The second way to use the ¹³C NMR spectrum is to check structural assignments that you have already made in working from the IR and proton NMR spectra.

From the infrared spectrum, you are usually able to identify the major functional groups within a molecule. Table 4.2 (on page 114) lists the ranges of ¹³C chemical shifts in which you would expect to see a peak for each of those groups. Thus, if you think that the compound in question contains an ester function and a *para*-disubstituted aromatic ring, you would expect to see a resonance between 155 and 175 ppm for the carbonyl group, and four peaks in the region between 110 and 135 ppm for the aromatic-ring carbon atoms. Furthermore, the peak assigned to the carbonyl carbon atom and two of the peaks assigned to phenyl carbon atoms should appear as singlets in the off-resonance decoupled spectrum, because none of those carbon atoms are

**Table 4.2**

Expected $^{13}$C resonances for confirmation of structural assignments based on IR and NMR spectra

| Functional or structural group | | Expected peak(s) in the broad-band decoupled $^{13}$C NMR spectrum (in ppm) | Expected appearance of peak(s) in the off-resonance $^{13}$C NMR spectrum[a] |
|---|---|---|---|
| Acid chloride (—CO—Cl) | | 1 peak, 185–165 | s |
| Alcohol (—C—OH) | 3° | 1 peak, 85–70 | s |
| | 2° | 1 peak, 75–60 | d |
| | 1° | 1 peak, 70–40 | t |
| Aldehyde (—CHO) | | 1 peak, 205–175 | d |
| Alkene (internal C=C) | | 2 peaks,[b] 150–110 | s or d only |
| Alkene (terminal C=C) | | 1 peak, 150–110 | s or d only |
| | | 1 peak, 150–110 | t |
| Alkyne (internal C≡C) | | 2 peaks,[b] 100–70 | s |
| Alkyne (terminal C≡C) | | 1 peak, 100–70 | s |
| | | 1 peak, 100–70 | d |
| Amide (—CO—NR$_2$) | | 1 peak, 180–160 | s |
| Amine (—C—NR$_2$) | | 1–3 peaks, 75–20 | s, d, t, or q |
| Anhydride (—CO—O—CO—) | | 2 peaks,[b] 175–150 | s |
| Benzenoid compounds | | 2–6 peaks,[b] 155–110 | s or d only |
| Carboxylic acid (—COOH) | | 1 peak, 185–160 | s |
| Ester (—COOR) | | 1 peak, 175–155 | s |
| Ether (—C—O—C—) | | 2 peaks,[b] 85–40 | s, d, t, or q |
| Ketone (—CO—) | | 1 peak, 225–175 | s |
| Nitrile (—C≡N) | | 1 peak, 130–110 | s |
| Methine group (—CH) | | varies, 60–30 | d |
| Methyl group (—CH$_3$) | | varies, 30–20 | q |
| Methylene group (—CH$_2$) | | varies, 45–25 | t |

[a] s = singlet, d = doublet, t = triplet, q = quartet.
[b] May observe only one peak because of symmetry or pseudosymmetry in the molecule.

attached to protons. The two remaining peaks in the aromatic region should be doublets in the off-resonance decoupled spectrum, because each has a proton attached.

Similarly, you should be able to confirm the proton NMR spectral assignments by observing peaks for the same C—H containing fragments. For example, if the proton NMR spectrum shows the presence of one methyl and two methylene groups, then the off-resonance decoupled $^{13}$C spectrum must have two triplets and one quartet at the appropriate chemical-shift values.

If an expected peak is not present, go back and reconsider your previous assignments, especially for functional groups containing the carbonyl moiety. A scarcity of peaks for the carbon skeleton may simply mean that two (or more) resonances happen to appear at the same chemical shift, or that a slow relaxation rate gives a resonance with very low intensity. The proton NMR will often resolve the ambiguity.

---

# References

R. J. Abraham and P. Loftus, *Proton and $^{13}$C NMR Spectroscopy: An Integrated Approach.* London: Heyden, 1978.

T. C. Farrar and E. D. Becker, *Pulse and Fourier Transform NMR.* New York: Academic Press, 1971.

G. C. Levy, R. L. Lichter, and G. L. Nelson, *Carbon-13 Nuclear Magnetic Resonance Spectroscopy.* New York: Wiley, 1980.

E. Pretsch, T. Clerc, J. Seibl, and W. Simon, *Tables of Spectral Data for Structure Determination of Organic Compounds.* Berlin: Springer-Verlag, 1983.

R. M. Silverstein, G. C. Bassler, and T. C. Morrill, *Spectrometric Identification of Organic Compounds.* New-York: Wiley, 4th ed., 1978.

# CHAPTER 5

# Mass Spectroscopy

The easiest way to identify an organic molecule would be to look at its whole structure at once, instead of focusing on the hydrogen or carbon atoms or on the functional groups separately. Of the many spectroscopic techniques available to the chemist, mass spectroscopy provides one of the few such structural probes of an entire molecule.

Mass spectroscopy is a relatively old analytical method, at least in concept; and as early as 1918, precise mass measurements were reported using an instrument similar to those in use today. As with every technique, mass spectroscopy has blossomed with the development of computer-interfaced instruments and with new ionization methods. However, because of the expense and effort needed to maintain mass spectrometers, they are still used largely by research groups or facilities that can employ them regularly. Thus, unlike the IR or NMR spectrum, recording a mass spectrum is not a routine procedure in a typical undergraduate laboratory.

However, mass spectroscopy is extremely useful as an analytical tool, and often you can obtain some of the necessary structural information only from the mass spectrum. Furthermore, for analyzing extremely small samples (picograms), or mixtures that can be separated by gas or liquid chromatography, mass spectroscopy is the only technique that will give *any* structural data.

## I. THEORY

In the gas phase, an organic molecule can be oxidized in a single-electron process to generate a cation-radical species. The principal concept underlying mass spectrometry is that the mass and charge of the species thus formed will cause it to travel a curved path in a magnetic field, according to Equation 5.1, where $m$ is the mass of the particle, $z$ is its charge, $H$ is the strength of the external magnetic field, $r$ is the radius of the path that the ion follows, and $V$ is the accelerating potential:

$$\frac{m}{z} = \frac{H^2 r^2}{2V}. \tag{5.1}$$

Placing a detector at a point to measure $r$, you would then be able to calculate the mass of the ion, if you know its charge. Although multiply charged organic molecules are formed in the gas phase, a charge of $+1$ is more usual.

Removal of an electron from the parent molecule by the process just described generates what is called the *molecular ion*, which is often stable enough to be detected. Since the mass spectrum is recorded on molecules in the gas phase at very low pressure, there are few collisions; so the molecular ion dissipates its energy by disintegration into smaller fragments. If the molecular ion or any of its fragments have lifetimes of at least $10^{-6}$ seconds, then they will reach the detector. By measuring the different values of $r$, or alternatively, by measuring the strength of the magnetic field necessary to cause an ion to travel a fixed radius, you will be able to calculate the mass of each fragment. Plotting the relative number of each ion reaching the detector (relative intensity) versus the mass of the ion (actually, mass/charge, $m/z$) generates the mass spectrum, Figure 5.1.

Before we look at the types of fragments that can form from the molecular ion, consider the mass spectrum shown in Figure 5.1. This type of spectrum is considered "low resolution," because we can differentiate only between fragments that differ by one mass unit. In a high-resolution spectrum, you can sometimes observe species with masses only several ten-thousandths apart. The peak at the highest $m/z$ values in a low-resolution spectrum often corresponds to the molecular ion, $M^{+\cdot}$; and because most atoms have isotopes, you will observe more than one peak at the highest $m/z$ values. In this example, the compound is 2-butanone, $C_4H_8O$. For carbon, the two principal isotopes are $^{12}C$ (98.89 percent) and $^{13}C$ (1.11 percent); for hydrogen, they are $^1H$ (99.98 percent) and $^2H$ (0.02 percent); and for oxygen, they are $^{16}O$ (99.76 percent), $^{17}O$ (0.04 percent), and $^{18}O$ (0.20 percent). Thus, the peak at $m/z$ 72 corresponds to the sum of the *principal* isotope masses, $^{12}C_4{}^1H_8{}^{16}O$ $((4 \times 12) + (8 \times 1) + (1 \times 16) = 72)$. Deciding which peak is $M^{+\cdot}$ will be covered in more detail in Section III.

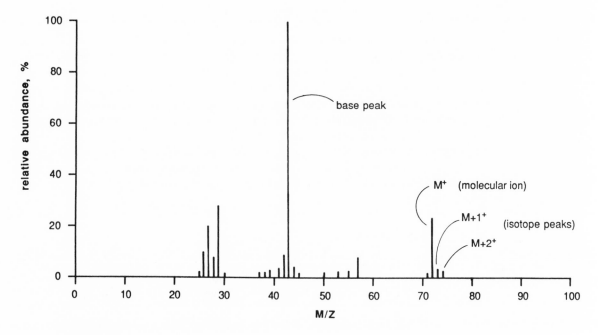

**Figure 5.1**
Mass spectrum of 2-butanone (MW 72.11) with significant peaks identified.

Another noticeable peak in the spectrum is the one with a relative intensity of 100 percent called the *base peak*. The base peak corresponds to the mass of an ion which is long-lived, and hence reaches the detector in greater quantity than any other. The base peak may sometimes be the molecular-ion peak, depending on the stability of $M^{+\cdot}$.

Except for the molecular-ion peaks, the remaining lines in the spectrum result from fragments. Although the process of fragmentation may seem random, there are actually only a few fragmentation pathways available to any one molecular ion. The stabilities of cationic species that you learned in an introductory organic chemistry course should suggest what kinds of species might be detected as "stable" fragments in a mass spectrometer; and if you already have an idea about the molecule's structure, you may be able to predict what type of fragments you should see.

Consider now the general process shown in Equation 5.2:

$$M \longrightarrow M^{+} \Big\langle \begin{array}{l} F_1 + F_2^{+} \longrightarrow \quad \cdots \cdot \\[2em] F_3 + F_4^{+} \longrightarrow F_5 + F_6^{+} \longrightarrow \quad \cdots \cdot \end{array} \tag{5.2}$$

What happens first is the loss of an electron from the molecule to form $M^{+\cdot}$, which is an odd-electron species (radical) carrying a positive charge. The radical can then fragment in two different ways, although at many different places within the molecule. The first pathway is by loss of a neutral molecule ($F_1$), leaving another cation-radical ($F_2^{+\cdot}$). The second pathway is by expulsion of a radical ($F_3^{\cdot}$), leaving a cation ($F_4^{+}$). Cation $F_4^{+}$ can then lose a neutral molecule ($F_5$), forming another cation ($F_6^{+}$). Both processes may continue until the species can no longer fragment, or until an especially stable cation is formed. Because species that are detected in the mass spectrum must carry a positive charge, only $M^{+\cdot}$, $F_2^{+\cdot}$, $F_4^{+}$, and $F_6^{+}$ in (5.2) will be detected.

As to detailed fragmentation patterns, many factors affect which bonds are broken. However, four types of cleavage are common to organic molecules. They are as follows.

### a. *Cleavage at branches*

Fragmentation of the hydrocarbon portion of a molecule occurs to form the more stable carbocation ($3° > 2° > 1°$). The largest possible group is lost as the radical portion, since the radical can be stabilized by delocalization through the $\sigma$-bonds.

$$(5.3)$$

### b. *α,β-Cleavage*

The bonds between the atoms that are $\alpha$ and $\beta$ to a heteroatom or $\pi$-bond undergo heterolytic cleavage, because the resulting cation is resonance-stabilized.

$$(5.4)$$

### c. *Loss of neutral molecules*

A cationic fragment, formed from $M^{+}$, may be cleaved further, losing an uncharged, stable molecule.

$$(5.5)$$

### d. *Rearrangements*

When the π-electrons are situated appropriately, substituents on the carbon frame-work can migrate to form rearranged species.

$$(5.6)$$

Rearrangements can be detected because, in their absence, a molecular ion with an even mass value cleaves to give fragment ions with odd-numbered masses, and vice versa. Thus, observation of fragments with the same parity (even from even or odd from odd) suggest that a rearrangement has taken place. A difference of one unit between the expected and observed mass for a fragment usually indicates that a hydrogen-atom migration has occurred.

## II. EXPERIMENTAL CONSIDERATIONS

The principal components of a "single focusing" mass spectrometer are sketched in Figure 5.2 (on page 122) and include an ionization source, an accelerator, a mass analyzer, an ion detector, and recorder. This type of instrument separates the ions by their behavior in a magnetic field, as discussed in Section I. The resolution is low in a single-focusing mass spectrometer, but operating such an instrument is fairly simple.

In the ionization source, the compound is introduced as a solid, liquid, or gas, with heating provided to vaporize nonvolatile compounds. Under vacuum, the resulting gaseous molecules then interact with an electron beam from a hot filament, in the simplest type of system, to generate ions. The typical energy used is 70 eV. For extremely nonvolatile, reactive, or labile substances, other ionization methods are available.

The ions thus formed are extracted by charged plates that accelerate the ions by repulsion at one plate and attraction at the other. Appropriately placed slits in the plates allow some of the ions to pass into the mass analyzer with sufficient kinetic energy to reach the detector. In a single-focusing instrument, the analyzer, also called a magnetic sector analyzer, is a curved tube within a magnetic field, in which the ions travel according to Equation 5.1. By changing the strength of the magnetic field, you can affect which ion travels the radius, $r$, at which the detector is located (Figure 5.2).

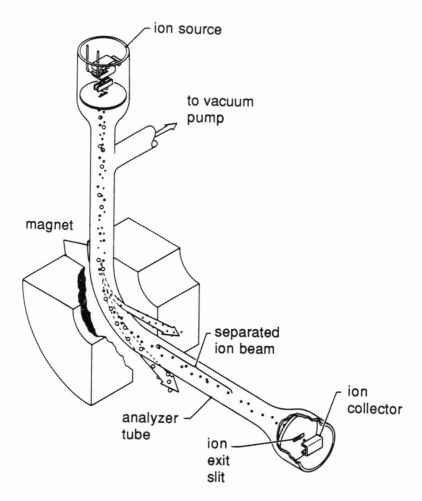

**Figure 5.2**
Diagram showing the components of a single-focusing mass spectrometer.

The detector measures the relative number of ions of a particular mass reaching it. By means of appropriate electronic amplification and recording, the detector signal is used to generate the spectrum.

(For more details about the actual workings of a mass spectrometer, sample introduction, and alternative ionization methods, see Chapman's text listed at the end of this chapter.)

# III. INTERPRETATION OF THE MASS SPECTRUM

Figure 1.3 (page 8) indicated that the mass spectrum is used at two points during the interpretation process. At the beginning, you will be trying to identify the molecular formula for the compound and which heteroatoms it contains. You may also be able to discover what general class of compound you have. At the end of the interpretation process, you can use the mass spectrum to confirm assignments made from interpretation of the other spectra.

## A. Initial Examination

### 1. Molecular Formula

When faced with the identification of an unknown substance, most chemists prefer to know the molecular formula of the compound. As noted in Chapter 1, you can use that knowledge to calculate the number of sites of unsaturation (pi-bonds and/or rings) in the molecule, and also to narrow the range of choices about the functional groups that may be present. Together, those data can often suggest what the structure of the unknown is, especially for simple compounds. Note, however, that you do *not* need to know the molecular formula in order to interpret the IR and NMR spectra.

To find the molecular formula from the mass spectrum, you first look for the peaks at the highest $m/z$ values. In the absence of other information, you can assume that one of the peaks at the highest $m/z$ values corresponds to the molecular ion. For many organic compounds, the molecular ion is observable; and peaks at $M + 1$ and $M + 2$ are also visible. Assuming Cl, Br, and/or S are *not* present (these heteroatoms are treated separately in the next section), then you will probably observe one of the four patterns shown in Figure 5.3 (on page 124). Pattern (a) shows a compound with readily discernible molecular-ion ($M^{+\cdot}$) and $M + 1$ peaks. In this instance, the $M + 2$ peak is very weak and is unobserved. In (b), $M^{+\cdot}$, $M + 1^{+\cdot}$, and $M + 2^{+\cdot}$ are all observed; so the identification of $M^{+\cdot}$ is straightforward. Finally in (c) and (d), *four* (or more) peaks are observed at the highest $m/z$ values. For these, since you will rarely observe an $M + 3$ peak, the peak at the highest $m/z$ value within the group results from the $M + 2^{+\cdot}$ fragment; therefore, $M^{+\cdot}$ is assigned as shown.

Another helpful guide (*not* a rule) is that many simple organic compounds have *even* molecular weights. If the most intense peak at the highest $m/z$ values occurs at an odd number, then that peak is either *not* $M^{+\cdot}$ or else the compound contains an odd number of nitrogen atoms. The presence of nitrogen in a molecule is discussed in more detail in the next section.

Finally, if you assign the peaks at highest $m/z$ values to $M^{+\cdot}$, yet observe peaks at $M - 3$ through $M - 13$, then you have probably not identified the actual molecular

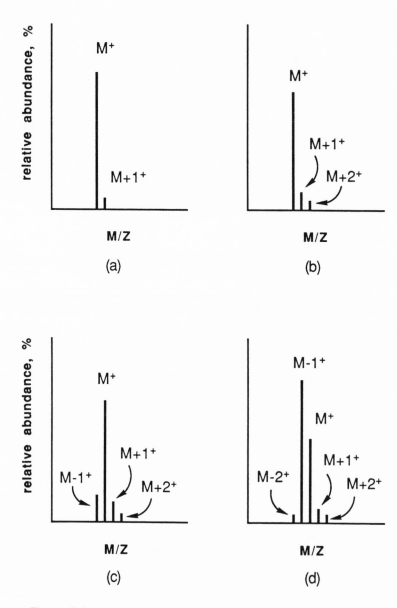

**Figure 5.3**
Typical patterns observed for the molecular ion region in mass spectra of organic compounds that do not contain S, Cl, or Br.

ion peak, or else the sample is contaminated. As you will see, formation of fragments from the parent molecule with masses between 3 and 13 less than the mass for the whole molecule is rare. Such peaks must either arise from impurities or be related to a compound with a larger mass (the actual molecular ion).

Sometimes the peaks at highest $m/z$ values do not correspond to the molecular ion or its isotopic cogeners. However, you will not necessarily discover this until you have started to examine the other spectra. Always proceed with caution, and be ready to revise your assignment of the molecular formula if new data conflict with your assignment made at this early stage.

Assuming you have identified the mass of the molecular ion, how do you deduce the molecular formula? As an example, let us assume that the mass of the molecular ion is $m/z$ 68. From the table in Appendix A, you can see there are three logical formulae having mass 68: $C_3H_4N_2$, $C_4H_4O$, and $C_5H_8$. The value 68 is what you calculate using the masses of the most abundant isotope for each element ($^1H$, $^{12}C$, $^{14}N$, and/or $^{16}O$). If you propose $C_4H_4O$ as the molecular formula, then the peak at $m/z$ 69 would correspond to a mixture of the following species:

$$(^{12}C)_3(^{13}C)(^1H)_4(^{16}O), \quad (^{12}C)_4(^1H)_3(^2H)(^{16}O), \quad (^{12}C)_4(^1H)_4(^{17}O) \quad (5.7)$$

You could calculate, using the known relative abundance for each of the different isotopes ($^1H$, $^2H$, $^{12}C$, $^{13}C$, $^{16}O$, and $^{17}O$), what the intensity of the $M + 1^{+\cdot}$ should be, compensating for the statistical distribution of the isotope for each element in the three species in (5.7). By measuring the actual intensity of the $M + 1^{+\cdot}$ peak in the mass spectrum, you could verify your assignment if the two values are similar. You could then carry out the same process for the $M + 2^+$ peak. Fortunately, you do not have to calculate the values each time, because detailed tables, like Table 5.1, have been compiled already by Beynon. Additional ones are collected in Appendix A and in Beynon's book, listed at the end of this chapter. The values given in the tables are only approximate, because the actual intensities of the isotope peaks can be affected by other factors, such as the presence of impurities or bimolecular collisions; however, they are often close enough to be helpful.

**Table 5.1**
Calculated intensity ratios for isotopic peaks of $M^{+\cdot} = 68$ $m/z$

| Formula | $M + 1^a$ | $M + 2^a$ |
|---|---|---|
| $C_3H_4N_2$ | 4.07 | 0.06 |
| $C_4H_4O$ | 4.43 | 0.28 |
| $C_5H_8$ | 5.53 | 0.12 |

[a] These numbers represent the relative intensity assuming that the intensity of $M^{+\cdot}$ is 100 percent.

There are other ways to find the molecular weights of organic compounds besides mass spectroscopy; these include freezing-point depression, boiling-point elevation, and vapor-phase equilibration. Therefore, if the mass spectrum is not available for an unknown compound that you are trying to identify, you can still get an idea about the molecular formula from laboratory measurements. The principal advantage of mass spectroscopy for finding the molecular weight is that very little material is needed to give an accurate mass value.

### 2. Heteroatoms

The initial examination of the mass spectrum is useful for finding out if the heteroatoms nitrogen, sulfur, bromine, and chlorine are present.

*Nitrogen.*   The presence of nitrogen in a molecule can be easily ascertained if there are an odd number of N atoms, because you will observe the molecular ion peak at a non-even $m/z$ value. This observation, known as the "nitrogen rule," results because a nitrogen atom requires one less substituent than carbon does to complete its valence. For an even number of nitrogen atoms, of course, the mass of $M^{+\cdot}$ will be even; so here you will have to rely on the intensities of $M^{+\cdot}$, $M + 1^{+\cdot}$, and $M + 2^{+\cdot}$ to find out whether the formula contains nitrogen. In any event, the functional group in most nitrogen-containing molecules ($C\equiv N$, $C=N$, $NO_2$, $NO$, $NH_2$, $NH$, amide) gives rise to readily observed bands in the infrared spectrum that can be used to verify the presence of nitrogen.

*Sulfur.*   The presence of sulfur likewise is usually discovered or verified by IR spectroscopy. In the mass spectrum, a sulfur atom manifests itself by making the $M + 2^{+\cdot}$ peak somewhat larger than expected. This pattern, illustrated in Figure 5.4(a), results because the natural abundance of $^{34}S$ is 4.22 percent compared to an abundance for $^{33}S$ of only 0.76 percent. Thus, the contribution of sulfur isotope to the intensity of the $M + 1^{+\cdot}$ peak is less than the contribution to $M + 2^{+\cdot}$. Because the $M + 2^{+\cdot}$ peak for most compounds is much smaller than the $M + 1^{+\cdot}$ peak, the presence of sulfur is readily detected by observing an $M + 2^{+\cdot}$ peak that is about 1/24 the intensity of $M^{+\cdot}$.

*Halogens.*   The presence of bromine and chlorine is also readily detected from the size of the $M + 2^{+\cdot}$ peak (Figure 5.4). For detecting the presence of the halogens, IR and NMR spectroscopy are often insensitive; so mass spectroscopy is essential for discovering their presence. For each, the $M + 2^{+\cdot}$ peak is so much more intense than it is for compounds lacking halogens that even a cursory glance at the mass spectrum reveals their presence. For chlorine, the relative amounts of $^{35}Cl$ and $^{37}Cl$ are 75.53 percent and 24.47 percent, respectively; and for bromine, the amounts of $^{79}Br$ and $^{81}Br$ are 50.54 percent and 49.46 percent, respectively. If two of these heteroatoms are present in a compound, then the effects are additive, and distinctive patterns are observed (Figure 5.5).

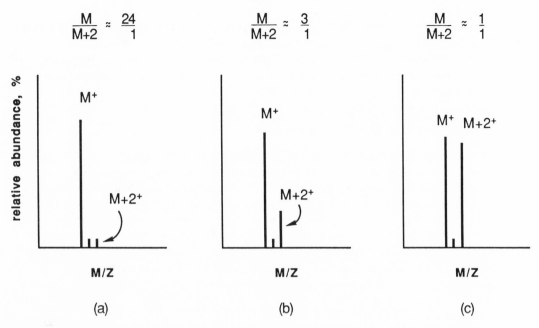

**Figure 5.4**
Appearance of the molecular ion region when the following heteroatom is present: (a) sulfur; (b) chlorine; and (c) bromine.

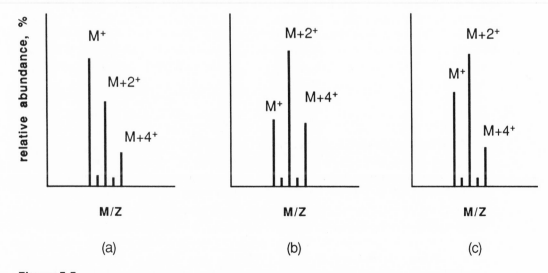

**Figure 5.5**
Appearance of the molecular ion region for a compound containing (a) two chlorine atoms; (b) two bromine atoms; and (c) one chlorine and 1 bromine atom.

Iodine is monoisotopic so cannot be detected by obvious differences in the intensities of the molecular-ion peaks, except that the observed heights for $M + 1^{+\cdot}$ and $M + 2^{+\cdot}$ will be smaller than expected from the mass of the compound. Because iodine is so massive (atomic weight = 126), it is sometimes easily detected by its loss during fragmentation.

### 3. Characteristic Fragments

If you wish to cull additional information from the mass spectrum before examining the IR and NMR spectra, look for a peak corresponding to a distinctive fragment of the molecule type. Without other data, any conclusion about the compound's structure based on only one or two peaks in the mass spectrum is risky. However, the loss of specific fragments from molecules in certain classes is quite predictable.

The first "fragment" is the molecular ion itself. Table 5.2 summarizes the expected intensities for various compound types, and those intensities can be used as a rough guide. Practically, you can use the intensity data only for the extremes; that is, if you observe a very intense molecular-ion peak, then you can probably eliminate alcohols from further consideration. You will never distinguish between an aromatic aldehyde and an aliphatic ketone simply from the intensity of $M^{+\cdot}$, however.

The next types of fragments worth noting include those that are unusually intense, for example, the base peak ion. Since the intense peaks often correspond to especially stable ions or radical-ions, they can often (but not always) form without a rearrangement taking place. Similarly, you can look for peaks that correspond to the loss of small fragments from the molecular ion. Table 5.3 lists such diagnostic fragments and the types of compounds that retain or lose such fragments. However, you must always be aware that rearrangements and other unexpected fragmentation processes can give misleading data. Every conclusion about a proposed structural component should be verified by an observation using another type of spectroscopy. A further discussion of the fragmentation pathways indicated in Table 5.3 are presented in the next section.

## B.  Confirmation of Structures

After you have proposed a structure based on the IR and NMR spectra, you can use the mass spectrum to confirm the presence of certain structural features.

### 1. Molecular Formula

If you already have a reasonable idea about the identity of the molecule, you should at least know the compound type (aromatic ketone, aliphatic nitrile, etc.). You can then use Table 5.2 to find the expected intensity for the molecular ion peak. For example, if you think that the compound is an aromatic ketone, then you know that the molecular ion should appear as an intense peak. If you see very small peaks at the

**Table 5.2**
Relative intensities of $M^{+\cdot}$ according to compound type

| Molecular-ion peak | Possible compounds |
|---|---|
| Prominent | Ar-X   X＝H, OH, NH$_2$, CO$_2$H, NO$_2$, CHO, ketone, halide, SH. |
| | Heteroaromatic compounds |
| | Alicyclic amines |
| | RSH |
| Medium Intensity | R-X   X＝H, —C≡C, CHO, CONH$_2$, ketone |
| | Ar-COOR |
| | RSR, RSSR |
| Weak Intensity | R-X   X＝CO$_2$H, NH$_2$, CN, CO$_2$R, OH (1° and 2°), Cl, OR′ |
| None | R-X   X＝OH (3°), NO$_2$, F, Br, I |

**Table 5.3**
Diagnostic fragments and associated classes of compounds[a]

| Mass | Fragment lost | Compound class |
|---|---|---|
| 1 | H | RCN, aldehydes, cyclopropyl compounds |
| 14 | N | Ar-NO |
| 15 | CH$_3$ | all types with an aliphatic portion |
| 16 | O | *N*-oxides, some sulfoxides |
| | NH$_2$ | amines |
| 17 | OH | carboxylic acids |
| 18 | H$_2$O | alcohols, higher MW aldehydes, ketones and ethers |
| 20 | HF | R-F |
| 27 | HCN | nitrogen heterocycles, nitriles, ArNH$_2$ |
| 28 | CO | aromatic oxy compounds (carbonyls, phenols), cyclic ketones |
| | CH$_2$＝CH$_2$ | cycloalkanes; aliphatic compounds >6 carbon atoms and having a $\pi$-bond |
| 30 | NO | ArNO$_2$ |
| 32 | S | sulfides, ArSH |
| | CH$_3$OH | methyl esters; methyl ethers |
| 34 | H$_2$S | thiols, methyl sulfides |
| 35 | Cl | R-Cl |
| 36 | HCl | R-Cl |
| 44 | CONH$_2$ | R-CONH$_2$ (stable R$^+$ only) |
| | CO$_2$ | carbonates, cyclic anhydrides |
| | CS | thiophenols, ArSAr |
| 46 | NO$_2$ | ArNO$_2$ |
| | CH$_2$S | ArSCH$_3$ |
| 64 | SO$_2$ | RSO$_2$R |

[a] R = alkyl, Ar = aryl.

highest $m/z$ values, then you may conclude such peaks actually result from an impurity in your sample. Once you have assigned the molecular ion, you can find the molecular formula as outlined earlier.

However, you may already know the empirical formula of your unknown before you examine its mass spectrum, because you can deduce the number of hydrogen and carbon atoms from the proton and $^{13}$C NMR spectra, respectively. Having proposed a structure, you should calculate the molecular weight, then see if the peaks at highest $m/z$ values are compatible with the formulation. The observed intensity of the peak you assign as $M^{+\cdot}$ must also be reasonable, as was mentioned.

Verification of the molecular formula is only the first step in using mass spectroscopy to check structural assignments based on other types of spectra. Many structures exist for any given molecular weight. Therefore, you must look for the expected fragmentation patterns for the assigned structure in order to make the best use of mass-spectral data.

### 2. Fragmentation Patterns

Section I presented four general modes of fragmentation that organic molecules undergo. If you have already proposed a structure for a compound, or even if you know only the major functional group from the IR spectrum, then you may be able to predict some fragments that will likely form or be lost from the parent molecule. Observation of peaks at $m/z$ values corresponding to the predicted values provides strong confirmatory evidence that the proposed structure is correct. The table in Appendix B lists the types of peaks that you can expect for different compounds, and the following discussion is arranged in the same order for easy reference. (More detailed discussions of molecules in each class are found in the book by Budzikiewicz, Djerassi, and Williams listed at the end of this chapter.)

#### a. *Hydrocarbons*

*Alkanes.*   You will rarely encounter unfunctionalized molecules if you continue to study organic chemistry; however, much of the early work in mass spectroscopy was performed on saturated hydrocarbons, mainly by petroleum chemists. Mass spectrometry is invaluable for identifying aliphatic hydrocarbons, because the proton NMR spectrum may not be resolved well enough to observe spin-spin splitting and to reveal the connectivity of the carbon chain.

The mass spectrum of a typical saturated aliphatic hydrocarbon, octane, at 70 eV follows the pattern illustrated in Figure 5.6. The fragmentation produces ions that give peaks separated by 14 mass units, corresponding to species differing by $CH_2$ groups. The intensities of the peaks can be traced with a curve shown by the dotted line. Although the molecular ion is usually observed, some of the higher molecular-weight fragments are sometimes absent. When the hydrocarbon chain is branched, then the smooth curve shows a discontinuity at the point where the chain is substituted. This situation, illustrated for 3-methyloctane in Figure 5.7, results because fragmentation at

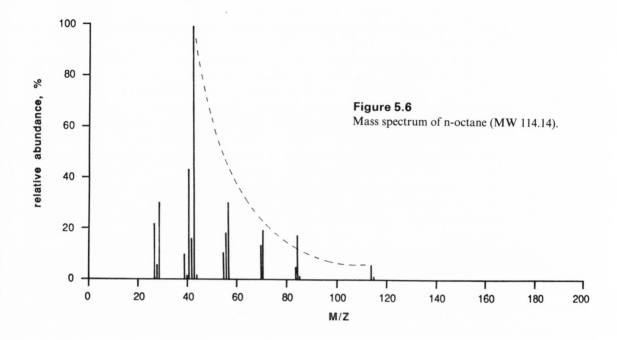

**Figure 5.6**
Mass spectrum of n-octane (MW 114.14).

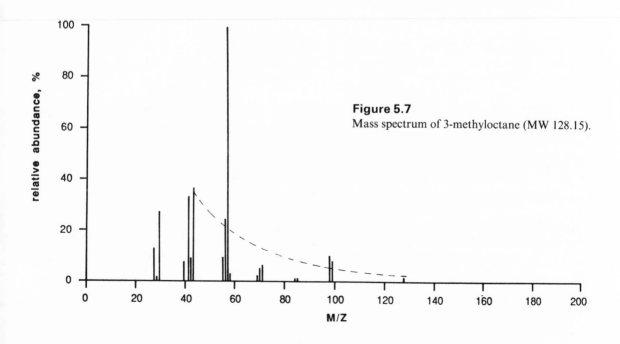

**Figure 5.7**
Mass spectrum of 3-methyloctane (MW 128.15).

a branch point gives rise to a more stable, hence longer-lived, secondary or tertiary carbocation.

For any aliphatic compound, the type of pattern observed for a simple alkane (Figure 5.6 or 5.7) will be superimposed on the patterns that result from any other functional group that may be present. Thus, a collection of peaks at 43, 57, 71, 85, etc., often indicates the presence of an alkyl chain in the molecule.

*Cycloalkanes.*   Cyclic aliphatic hydrocarbons have one site of unsaturation; therefore the series of peaks resulting from fragmentation often occur at 2 mass units less than the peaks for an acyclic compound, that is, at 41, 55, 69, etc. The molecular ion is often more pronounced in intensity than for a straight chain compound; and cleavage of side chains and substituents occurs readily, for the same reason as for branched acyclic compounds. A major pathway for fragmentation of cyclic compounds is by loss of $CH_2{=}CH_2$ ($m/z$ 28), resulting in a greater proportion of even-mass fragments than for acyclic hydrocarbons. Figure 5.8 shows the mass spectrum of ethylcyclohexane.

*Alkenes.*   Like cycloalkanes, alkenes have one site of unsaturation. Thus, their mass spectra show patterns with peaks at 41, 55, 69, etc., since fragmentation of the saturated part of the molecule occurs by the same process as fragmentation of alkanes. Allylic cleavage is particularly favored because the resulting cation is resonance stabilized. In addition, rearrangements are also possible:

$$(5.8)$$

You cannot usually identify the position of the double bond by mass spectroscopy since the pi-bond can migrate during the ionization or fragmentation processes. The mass spectrum of 2-octene is shown in Figure 5.9.

*Arenes.*   Aromatic hydrocarbons are different from their aliphatic counterparts, in that the molecular-ion peak is more intense. Aromatic rings having an alkyl substituent give a particularly characteristic ion that results from benzylic cleavage and rearrangement to form the so-called tropylium cation (see Figure 5.10, page 134). Additional substitution of the ring will still result in formation of the benzyl and tropylium cations, except that their masses will be increased by the mass of the substituent. The tropylium cation can subsequently fragment by losing acetylene to give a species with $m/z$ 65.

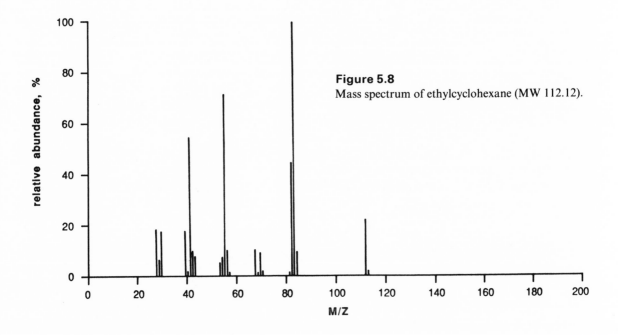

**Figure 5.8**
Mass spectrum of ethylcyclohexane (MW 112.12).

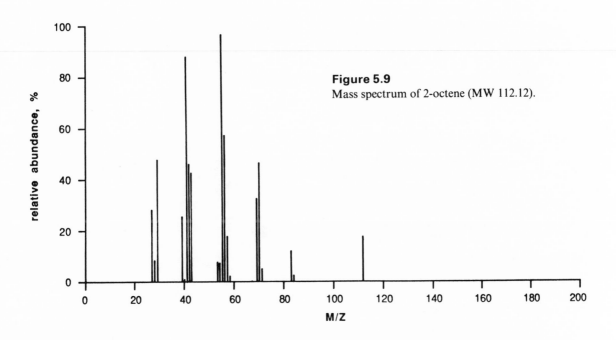

**Figure 5.9**
Mass spectrum of 2-octene (MW 112.12).

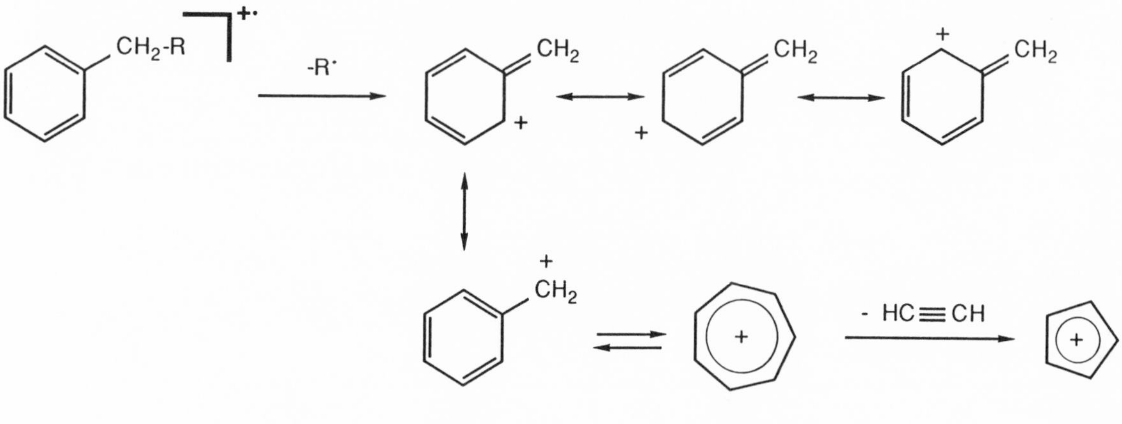


M/Z 91 M/Z 91 M/Z 65

**Figure 5.10**
Typical fragmentation pathways observed for an alkyl-substituted benzene compound.

Arene derivatives also fragment at the ring (see 5.9) to give the corresponding phenyl cation ($m/z$ 77 when Ar = phenyl), and/or at the $\beta$-carbon (see 5.10) with concomitant H-migration ($m/z$ 92 when Ar = phenyl):

$$\text{Ar-R}^{+\cdot} \longrightarrow \underset{\text{M/Z 77}}{\text{Ar}^{+}} + \text{R}^{\cdot} \qquad (5.9)$$

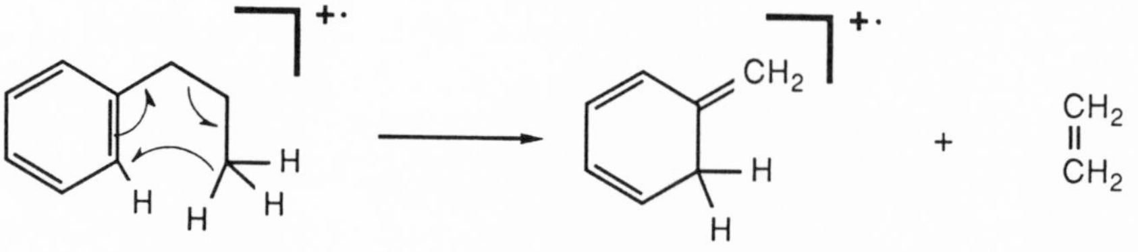


(5.10)

M/Z 92

You should look for all three types of fragments if you think the unknown is a substituted benzene compound.

### b. *Halocarbons*

You have already seen that the presence of bromine or chlorine is signaled by the appearance of two isotope peaks in the molecular-ion region. For aliphatic chlorides and bromides, the molecular-ion peak is often weak; and the fragmentation patterns are very much like the patterns for the structurally analogous hydrocarbons. Of course, any fragment with mass $N$, containing the halogen atom, will show the characteristic isotope pattern at $m/z\ N$ and $m/z\ (N + 2)$. Intense peaks corresponding to the ions $C_4H_8Cl^+$ and $C_4H_8Br^+$ are observed for straight-chain halocarbons with more than six carbon atoms; these ions are likely cyclic species having the structure shown in (5.11):

$$
\begin{array}{ccc}
\begin{array}{c}
\text{H}_2\text{C} - \text{CH}_2 \\
/ \qquad \backslash \\
\text{H}_2\text{C} \qquad \text{CH}_2 \\
\backslash \quad \text{Cl} \quad / \\
+ \\
\text{M/Z 91}
\end{array}
&
\begin{array}{c}
\text{H}_2\text{C} - \text{CH}_2 \\
/ \qquad \backslash \\
\text{H}_2\text{C} \qquad \text{CH}_2 \\
\backslash \quad \text{Br} \quad / \\
+ \\
\text{M/Z 135}
\end{array}
& \qquad (5.11)
\end{array}
$$

The mass spectrum of 1-bromohexane is shown in Figure 5.11.

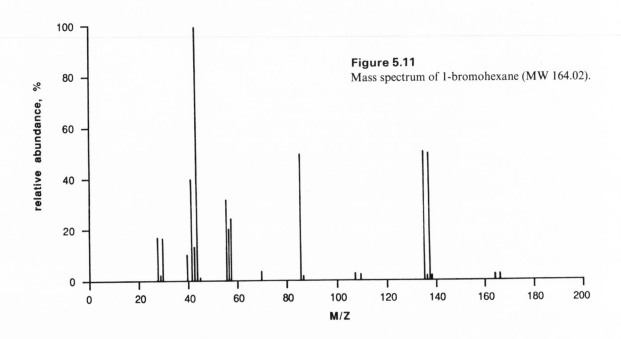

**Figure 5.11**
Mass spectrum of 1-bromohexane (MW 164.02).

Aliphatic iodides give the most intense molecular-ion peaks among halohydro-carbons. The presence of iodine is usually apparent, because the number of hydrogen and carbon atoms deduced from the NMR spectra seems too low in relation to the observed molecular weight. Additionally, you can sometimes detect the presence of iodine by observing large gaps between $M^{+\cdot}$ and fragment-ion peaks.

The aryl halides do not suffer from the same disadvantage of weak $M^{+\cdot}$ peaks. Usually, the molecular-ion peak as well as the $(M - X)^+$ peaks are moderately intense.

### c. Hydroxy compounds

*Alcohols.* The molecular-ion peak of alcohols is usually weak, and for 3° alcohols may be nonexistent. The $M^{+\cdot}$ region is sometimes further complicated, because the peaks at highest $m/z$ values actually correspond to $M - 2$ ($R—CH{=}O^{+\cdot}$) or even the rare $M - 3$ ($R—C{\equiv}O^+$) fragments. Another possibly prominent peak results from loss of water, and the $M - 18$ peak may appear at the highest $m/z$ value.

The fragmentation of alcohols occurs primarily at the $C—C$ bond $\alpha$ to the oxygen atom, and results in a prominent peak at $m/z$ 31 for a primary alcohol, at $m/z$ $\{31 + (R)\}$ [45, 59, 73, etc.] for a secondary alcohol, or at $m/z$ $\{31 + (R) + (R')\}$ [59, 73, etc.] for a tertiary alcohol:

$$\tag{5.12}$$

The remainder of the fragmentation pattern mimics the pattern of the correspond-ing hydrocarbon except that the peaks occur at 45, 59, 73, etc., because oxygen has a mass of 16 compared with the mass for a $CH_2$ group of 14. In addition, the normal hydrocarbon pattern is superimposed on the pattern of the oxygen-containing frag-ments. Finally, peaks resulting from elimination of both water and methyl ($M - 33$) or of water and alkene ($M - 46$, $M - 74$, $M - 102$, etc.) are also observed. The loss of water and alkene occurs via a cyclic transition state:

$$\tag{5.13}$$

The mass spectrum of 3-hexanol (Figure 5.12) summarizes the foregoing points.

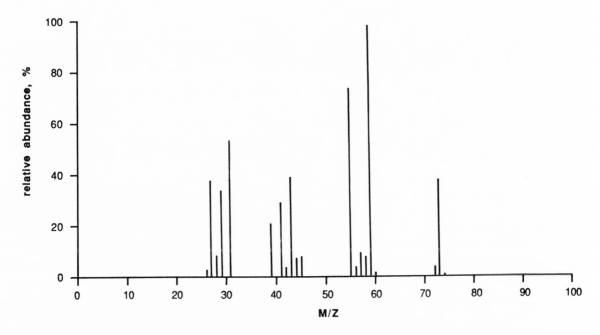

**Figure 5.12**
Mass spectrum of 3-hexanol (MW 102.10).

*Phenols.*    Phenols generally display a strong molecular ion peak and a variable intensity $M - 1$ peak. The latter is enhanced when there is an alkyl group on the aromatic ring and the benzyl/tropylium cation can form. Prominent peaks at $M - 28$ and $M - 29$ indicate the loss of CO and the CHO fragment, respectively:

(5.14)

M/Z 66                M/Z 65

Figure 5.13 shows the mass spectrum of 3-ethylphenol.

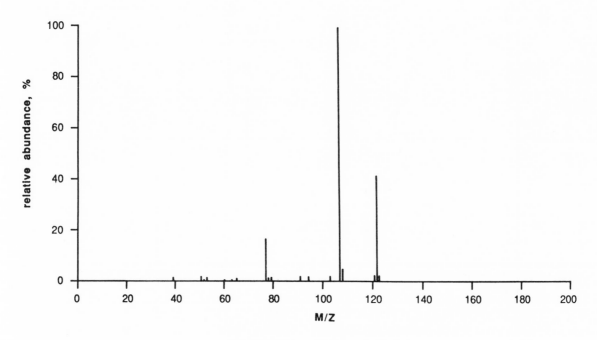

**Figure 5.13**
Mass spectrum of 3-ethylphenol (MW 122.07).

### d. *Ethers*

Aliphatic ethers (R—O—R′) often give small but discernible molecular ion peaks. Cleavage occurs next to the oxygen atom and at the α C—C bond, leading to $RO^+$ and $ROCH_2{}^+$ fragments at 31, 45, 59, 73, . . . . and $R^+$ fragments at 29, 43, 57, . . . . During the C—C bond cleavage pathway, the fragment that forms first may further expel an alkene. Dibutylether, whose spectrum is shown in Figure 5.14, illustrates the different types of fragmentation for aliphatic ethers (Figure 5.15).

Aromatic ethers have prominent molecular-ion peaks, and their fragmentation patterns in the mass spectrum are directed by the presence of the aromatic ring. The major pathway for cleavage in molecules of the form Ar—O—R involves breaking the C—O bonds, forming the aryl and aryloxy cations. If the side chain is anything

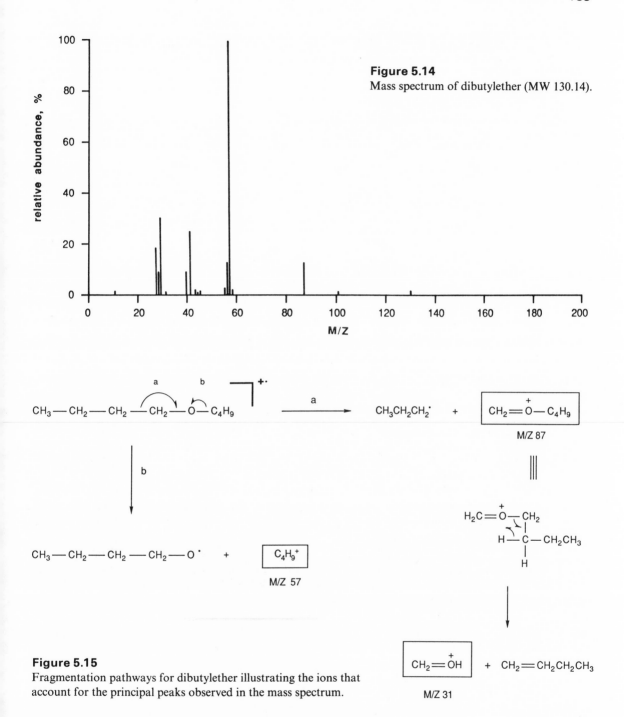

**Figure 5.14**
Mass spectrum of dibutylether (MW 130.14).

**Figure 5.15**
Fragmentation pathways for dibutylether illustrating the ions that account for the principal peaks observed in the mass spectrum.

longer than a methyl group, H-atom migration can also occur. Unlike dialkylethers, C—C bond cleavage α to the oxygen atom is unimportant:

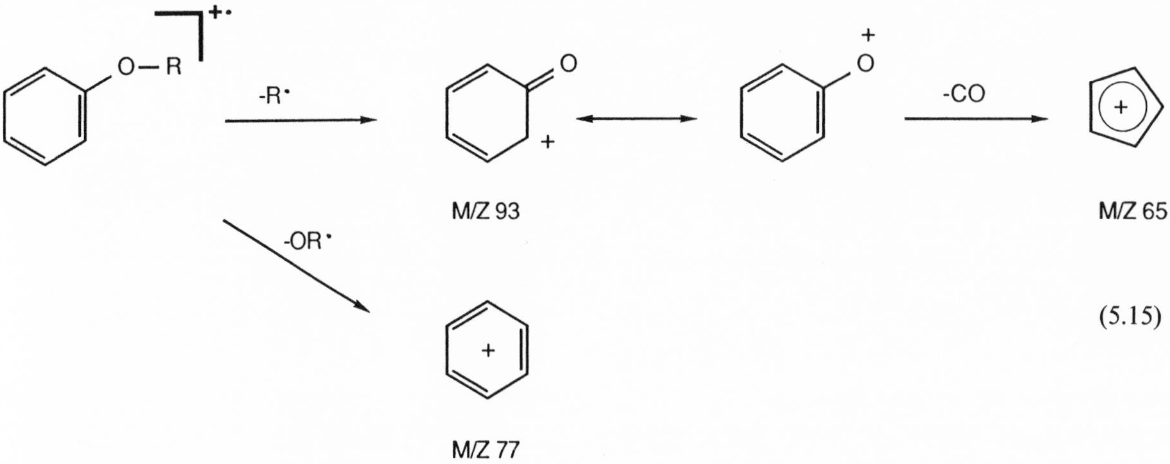

M/Z 93

M/Z 65

-OR·

M/Z 77

(5.15)

### e. *Aldehydes*

Aliphatic aldehydes have observable, but weak, molecular-ion peaks. Cleavage occurs primarily next to the carbonyl group, leading to fragments at $M - 1$ (loss of H) and $M - 29$ (loss of CHO). For compounds with more than four carbons, the "McLafferty rearrangement" generates a fragment with a mass of 43 + (R), where (R) is the mass of R in the structure in Figure 5.16. The mass spectrum of 3-methylpentanal is shown in Figure 5.17.

Aromatic aldehydes have moderate to strong molecular-ion peaks. Rupture of the C—H bond at the carbonyl group is the dominant fragmentation pathway, and the $M - 1$ peak is often larger than the $M^{+\cdot}$ peak. The first-formed $ArC\equiv O^+$ ion will lose CO to give $Ar^+$ ($m/z$ 77 for phenyl).

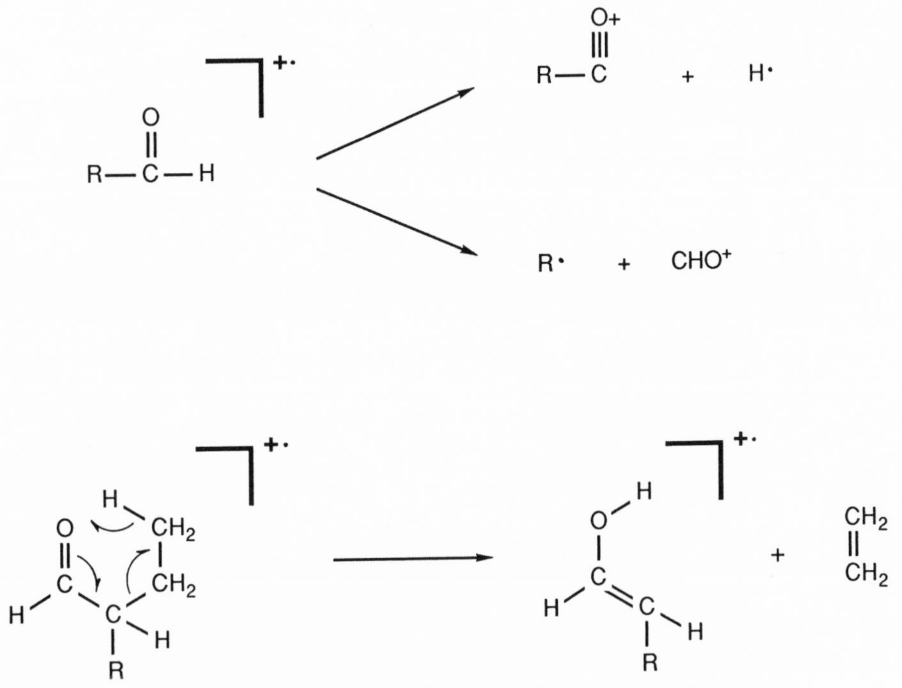

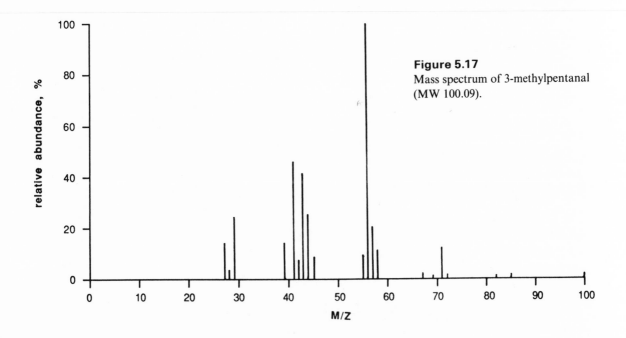

**Figure 5.16**
Fragmentation pathways of aldehydes.

**Figure 5.17**
Mass spectrum of 3-methylpentanal (MW 100.09).

### f. Ketones

The molecular-ion peak for an aliphatic ketone is usually prominent. Fragmentation occurs by cleavage of the C — C bond attached to the carbonyl carbon atom, and results in the formation of the stabilized $R—C\equiv O^+$ and $R'—C\equiv O^+$ ions, in an asymmetric ketone. When one of the chains is longer than two carbon atoms, the McLafferty cleavage becomes an important pathway, leading to peaks at $m/z$ 58 or 72 or 86, ... depending on the nature of R and R' in Equation (5.16). The spectrum of 3-heptanone is shown in Figure 5.18.

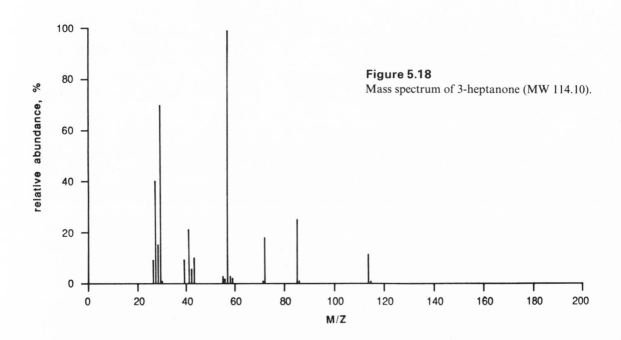

$$\text{(5.16)}$$

Aromatic ketones also give strong molecular-ion peaks, and the fragmentation process is the same as for the aliphatic case. For alkyl aryl ketones, the preferred cleavage occurs on the alkyl side, giving $Ar—C\equiv O^+$, which can lose CO to form $Ar^+$ ($m/z$ 77 for phenyl). Diaryl ketones give both $Ar—C\equiv O^+$ and $Ar'—C\equiv O^+$ fragments.

**Figure 5.18**
Mass spectrum of 3-heptanone (MW 114.10).

### g. *Carboxylic acids and esters*

Like the other types of carbonyl compounds, acids and esters fragment by cleavage of the bonds $\alpha$ to the carbonyl group, including both $C-C$ and $C-O$ linkages. The molecular-ion peak for aliphatic acids is usually weak; but for esters, $M^{+\cdot}$ is somewhat stronger.

The McLafferty rearrangement is facile for both acids and esters, and is usually very prominent. For many of the straight-chain acids and esters, in fact, the rearrangement process accounts for the base peak ion:

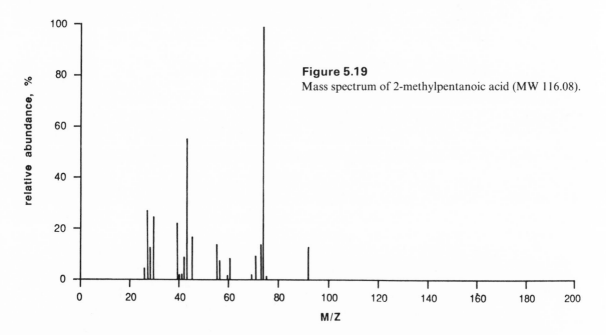

$$(5.17)$$

R = H, alkyl

Peaks at $M - 17$ (loss of OH) and $M - 45$ (loss of COOH) are important for acids; and peaks at $M - 31$ (loss of $OCH_3$) and $M - 59$ (loss of $COOCH_3$) are observed for many methyl esters. Corresponding homologous esters display loss of the appropriate fragments at multiples of 14 mass units more. Mass spectra of 2-methylpentanoic acid and methyl pentanoate are shown in Figures 5.19 and 5.20 (on page 144), respectively.

**Figure 5.19**
Mass spectrum of 2-methylpentanoic acid (MW 116.08).

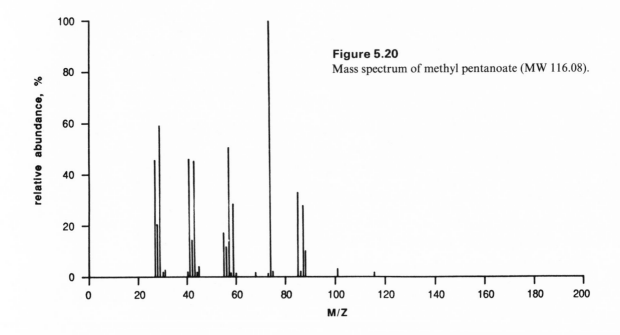

**Figure 5.20**
Mass spectrum of methyl pentanoate (MW 116.08).

Aromatic acids and esters (at least, methyl and ethyl esters) show prominent molecular-ion peaks. Fragmentation patterns observed for the aliphatic analogs hold for the aromatic compounds, too. Thus peaks at $M - 17$ and $M - 45$ for acids, and $M - 31$ and $M - 59$ for methyl esters are observed. Additionally, a peak at $M - 18$ for acids and at $M - 32$ for methyl esters result from elimination of $H_2O$ or $CH_3OH$ by the "ortho effect," in which a six-membered transition state can form:

(5.18)

$X = CH_2, O, NH$

### h. *Amines*

The molecular-ion peak of aliphatic amines is weak to unobservable, and occurs at a non-even $m/z$ value. The primary cleavage occurs, as in alcohols, at the $C-C$ bond $\alpha$ to the nitrogen atom unless the carbon chain is branched at the carbon atom bonded to N. The cleavage usually accounts for the base peak; for primary amines, it gives a peak with $m/z$ 30 ($CH_2NH_2^+$):

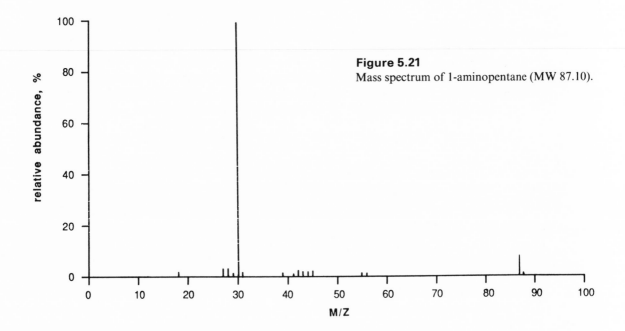

(5.19)

Because of the presence of the nitrogen atom, the fragmentation pattern for an aliphatic amine consists of peaks at *even* $m/z$ values, that is 30, 44, 58 . . . instead of the customary 29, 43, 57 . . . values observed for hydrocarbons. However, the two series are superimposed on each other, resulting in characteristic clusters of peaks. The mass spectrum of 1-aminopentane is shown in Figure 5.21.

**Figure 5.21**
Mass spectrum of 1-aminopentane (MW 87.10).

Aromatic 1°-amines (anilines) give intense molecular ion peaks at non-even $m/z$ values. Loss of a hydrogen atom from the amino group gives a moderately intense peak at $M - 1$, and loss of HCN results in a fragment with a mass of 66 for aniline itself. $N$-Alkylanilines cleave at the $C-C$ bond adjacent to nitrogen in the alkyl chain (5.20), in contrast to the behavior of alkylarylethers, in which $C-C$ bond rupture is *not* observed (5.21):

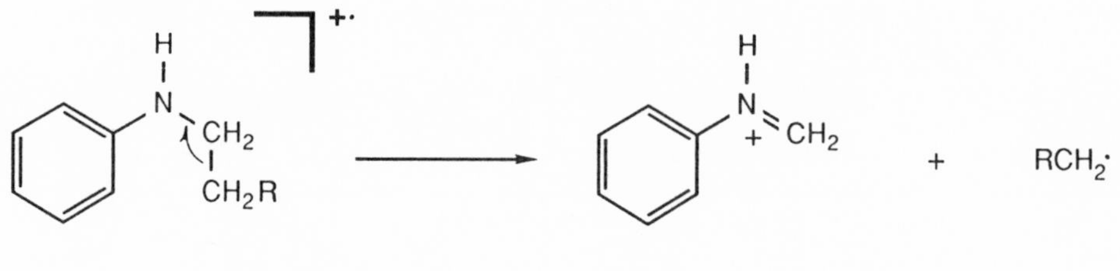

(5.20)

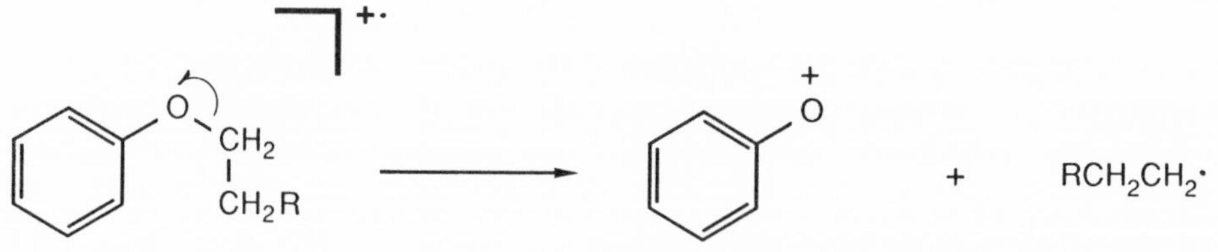

(5.21)

### i. Amides

Aliphatic amides display a weak but discernible molecular-ion peak. Their fragmentation follows the patterns seen for other carbonyl compounds, namely, the McLafferty rearrangement and cleavage of the $C-C$ bond attached to the carbonyl carbon. These cleavage patterns for butanoamide are:

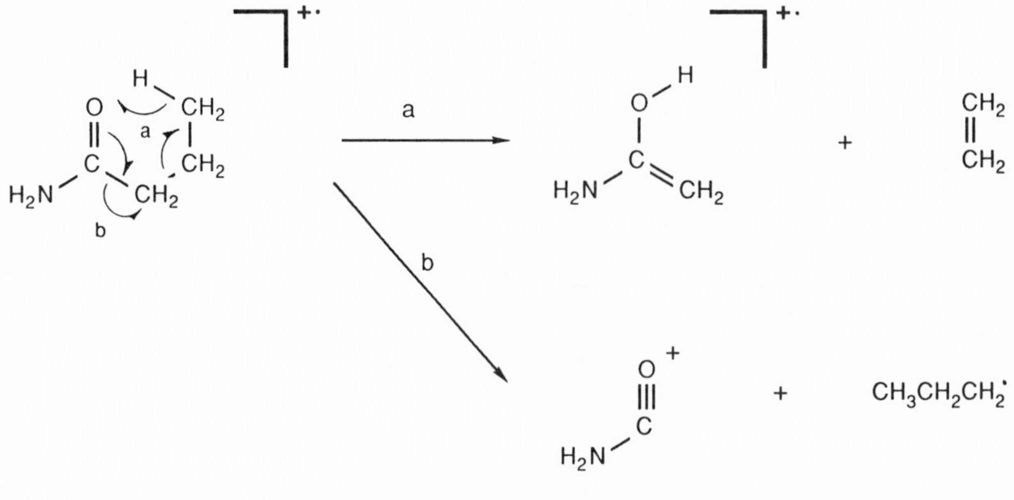

(5.22)

In addition, acetamides undergo a double bond fission to generate the $CH_2NH_2^+$ fragment at $m/z$ 30:

$$(5.23)$$

### j. *Nitriles*

The molecular-ion peak for aliphatic nitriles is often absent, although an $M - 1$ peak, resulting from loss of a hydrogen atom, can sometimes be seen:

$$(5.24)$$

Since the IR spectrum will usually provide evidence for the presence of the nitrile group, you should not be led astray by assigning the $M - 1$ peak as $M^{+\cdot}$ because you will know that the molecular weight should be an odd number. The dominant fragmentation pathway for aliphatic nitriles is by the McLafferty rearrangement, which results in a peak at $m/z$ 41:

$$(5.25)$$

The mass spectrum of pentanonitrile is shown in Figure 5.22.

Aromatic nitriles have prominent molecular-ion peaks, and the elimination of HCN is important ($m/z$ 76 for benzonitrile) in the fragmentation process.

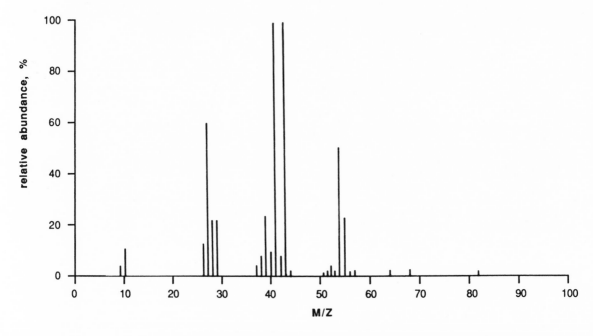

**Figure 5.22**
Mass spectrum of pentanonitrile (MW 83.07).

### k. *Nitro compounds*

Aromatic nitro compounds, which are encountered more often than the aliphatic analogs, have prominent molecular-ion peaks. Fragmentation occurs at the ring to give strong peaks at $M - 46$ (loss of $NO_2$) and $M - 30$ (loss of NO with rearrangement). Both of these first-formed cations can fragment further to give diagnostic peaks at $M - 72$ and $M - 58$ (see Figure 5.23). The mass spectrum of nitrobenzene is shown in Figure 5.24.

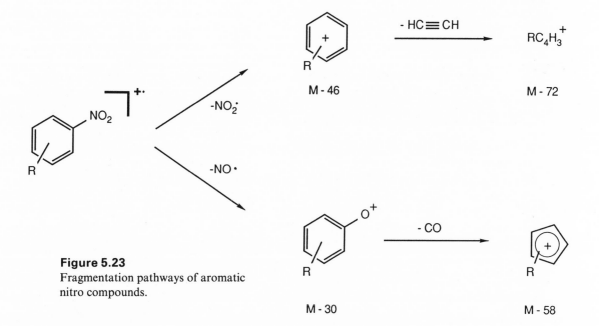

**Figure 5.23**
Fragmentation pathways of aromatic nitro compounds.

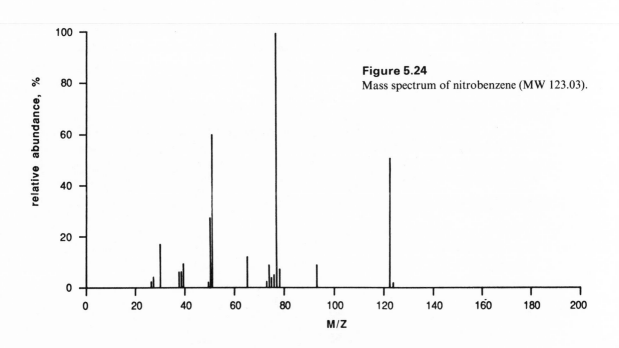

**Figure 5.24**
Mass spectrum of nitrobenzene (MW 123.03).

### I. *Thiols and sulfides*

Thiols, unlike alcohols, usually display a molecular-ion peak intense enough that the $M + 2^{+\cdot}$ peak can be measured easily. The fragmentation patterns are much like those of alcohols, except that the resulting peaks reflect the higher mass of S relative to O. Thus, a peak at $m/z$ 47 for the $CH_2=SH^+$ ion is indicative of a primary mercaptan, and a strong peak at $M - 34$ is evidence for the loss of $H_2S$. The mass spectrum of isobutylmercaptan is shown in Figure 5.25.

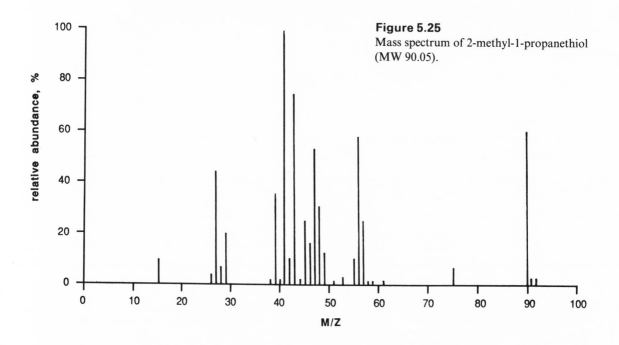

**Figure 5.25**
Mass spectrum of 2-methyl-1-propanethiol (MW 90.05).

Sulfides have detectable molecular-ion peaks, and their cleavage patterns resemble the patterns observed for ethers. Thus, $R—S—R'$ cleaves to give $RS^+$ and $R'S^+$ fragments. The ethylthio and pentylthio cations are especially stabilized, apparently by formation of cyclic species:

$$(5.26)$$

M/Z 61          M/Z 103

# References

J. H. Beynon, *Mass Spectrometry and Its Applications to Organic Chemistry*. Amsterdam: Elsevier, 1960.

H. Budzikiewicz, C. Djerassi, and D. H. Williams, *Mass Spectrometry of Organic Compounds*. San Francisco: Holden-Day, 1967.

J. R. Chapman, *Practical Organic Mass Spectrometry*. New York: Wiley, 1985.

F. W. McLafferty, *Interpretation of Mass Spectra*. Mill Valley, CA: University Science Books, 1983.

R. M. Silverstein, G. C. Bassler, and T. C. Morrill, *Spectrometric Identification of Organic Compounds*. New York: Wiley, 4th ed., 1978.

# Masses and Isotopic Abundance Ratios for Various Combinations of Carbon, Hydrogen, Nitrogen, and Oxygen*

| | $M+1$ | $M+2$ | | $M+1$ | $M+2$ | | $M+1$ | $M+2$ |
|---|---|---|---|---|---|---|---|---|
| **40** | | | **50** | | | **59** | | |
| $C_3H_4$ | 3.31 | 0.04 | $C_4H_2$ | 4.34 | 0.07 | $CHNO_2$ | 1.56 | 0.41 |
| | | | | | | $C_2H_5NO$ | 2.66 | 0.22 |
| **41** | | | **52** | | | $C_3H_9N$ | 3.77 | 0.05 |
| $C_2H_3N$ | 2.59 | 0.02 | $C_2N_2$ | 2.92 | 0.03 | | | |
| | | | $C_4H_4$ | 4.39 | 0.07 | **60** | | |
| **42** | | | | | | $CH_4N_2O$ | 1.95 | 0.21 |
| $CH_2N_2$ | 1.88 | 0.01 | **53** | | | $C_2H_4O_2$ | 2.30 | 0.04 |
| $C_2H_2O$ | 2.23 | 0.21 | $C_3H_3N$ | 3.67 | 0.05 | $C_2H_8N_2$ | 3.05 | 0.03 |
| $C_3H_6$ | 3.34 | 0.04 | | | | $C_3H_8O$ | 3.41 | 0.24 |
| | | | **54** | | | | | |
| **43** | | | $C_2H_2N_2$ | 2.96 | 0.03 | **61** | | |
| $C_2H_5N$ | 2.62 | 0.02 | $C_3H_2O$ | 3.31 | 0.24 | $CH_3NO_2$ | 1.59 | 0.41 |
| | | | $C_4H_6$ | 4.42 | 0.07 | $C_2H_7NO$ | 2.69 | 0.22 |
| **44** | | | | | | | | |
| $CH_4N_2$ | 1.91 | 0.01 | **55** | | | **62** | | |
| $C_2H_4O$ | 2.26 | 0.21 | $C_2HNO$ | 2.60 | 0.22 | $CH_6N_2O$ | 1.98 | 0.21 |
| $C_3H_8$ | 3.37 | 0.04 | $C_3H_5N$ | 3.70 | 0.05 | $C_2H_6O_2$ | 2.34 | 0.42 |
| | | | | | | $C_5H_2$ | 5.44 | 0.12 |
| **45** | | | **56** | | | | | |
| $CH_3NO$ | 1.55 | 0.21 | $C_2H_4N_2$ | 2.99 | 0.03 | **63** | | |
| $C_2H_7N$ | 2.66 | 0.02 | $C_3H_4O$ | 3.35 | 0.24 | $CH_5NO_2$ | 1.62 | 0.41 |
| | | | $C_4H_8$ | 4.45 | 0.08 | | | |
| **46** | | | | | | **64** | | |
| $CH_2O_2$ | 1.19 | 0.40 | **57** | | | $CH_4O_3$ | 1.26 | 0.60 |
| $CH_6N_2$ | 1.94 | 0.01 | $C_2H_3NO$ | 2.63 | 0.22 | $C_5H_4$ | 5.47 | 0.12 |
| $C_2H_6O$ | 2.30 | 0.22 | $C_3H_7N$ | 3.74 | 0.05 | | | |
| | | | | | | **65** | | |
| **47** | | | **58** | | | $C_4H_3N$ | 4.75 | 0.09 |
| $CH_5NO$ | 1.58 | 0.21 | $CH_2N_2O$ | 1.92 | 0.21 | | | |
| | | | $C_2H_2O_2$ | 2.27 | 0.42 | **66** | | |
| **48** | | | $C_2H_6N_2$ | 3.02 | 0.03 | $C_3H_2N_2$ | 4.04 | 0.06 |
| $CH_4O_2$ | 1.22 | 0.40 | $C_3H_6O$ | 3.38 | 0.24 | $C_4H_2O$ | 4.39 | 0.27 |
| | | | $C_4H_{10}$ | 4.48 | 0.08 | $C_5H_6$ | 5.50 | 0.12 |

**153**

* Adapted with permission from J. H. Beynon, *Mass Spectrometry and Its Application to Organic Chemistry*, Elsevier, Amsterdam, 1960. The values given form a self-consistent set and can be used regardless of the mass standard.

| | $M+1$ | $M+2$ |
|---|---|---|
| **67** | | |
| $C_4H_5N$ | 4.78 | 0.09 |
| | | |
| **68** | | |
| $C_3H_4N_2$ | 4.07 | 0.06 |
| $C_4H_4O$ | 4.43 | 0.28 |
| $C_5H_8$ | 5.53 | 0.12 |
| | | |
| **69** | | |
| $C_3H_3NO$ | 3.71 | 0.25 |
| $C_4H_7N$ | 4.82 | 0.09 |
| | | |
| **70** | | |
| $C_2H_2N_2O$ | 3.00 | 0.23 |
| $C_3H_2O_2$ | 3.35 | 0.44 |
| $C_3H_6N_2$ | 4.10 | 0.07 |
| $C_4H_6O$ | 4.46 | 0.28 |
| $C_5H_{10}$ | 5.56 | 0.13 |
| | | |
| **71** | | |
| $C_3H_5N_3$ | 3.39 | 0.04 |
| $C_3H_5NO$ | 3.74 | 0.25 |
| $C_4H_9N$ | 4.85 | 0.09 |
| | | |
| **72** | | |
| $C_2H_4N_2O$ | 3.03 | 0.23 |
| $C_3H_4O_2$ | 3.38 | 0.44 |
| $C_3H_8N_2$ | 4.13 | 0.07 |
| $C_4H_8O$ | 4.49 | 0.28 |
| $C_5H_{12}$ | 5.60 | 0.13 |
| | | |
| **73** | | |
| $C_2H_3NO_2$ | 2.67 | 0.42 |
| $C_3H_7NO$ | 3.77 | 0.25 |
| $C_4H_{11}N$ | 4.88 | 0.10 |
| | | |
| **74** | | |
| $CH_2N_2O_2$ | 1.95 | 0.41 |
| $C_2H_2O_3$ | 2.31 | 0.62 |
| $C_2H_6N_2O$ | 3.06 | 0.23 |
| $C_3H_6O_2$ | 3.42 | 0.44 |
| $C_3H_{10}N_2$ | 4.17 | 0.07 |
| $C_4H_{10}O$ | 4.52 | 0.28 |
| $C_6H_2$ | 6.52 | 0.18 |
| | | |
| **75** | | |
| $C_2H_5NO_2$ | 2.70 | 0.43 |
| $C_3H_9NO$ | 3.81 | 0.25 |

| | $M+1$ | $M+2$ |
|---|---|---|
| **76** | | |
| $CH_4N_2O_2$ | 1.99 | 0.41 |
| $CH_8N_4$ | 2.73 | 0.03 |
| $C_2H_4O_3$ | 2.34 | 0.62 |
| $C_2H_8N_2O$ | 3.09 | 0.24 |
| $C_3H_8O_2$ | 3.45 | 0.44 |
| $C_6H_4$ | 6.55 | 0.18 |
| | | |
| **77** | | |
| $CH_3NO_3$ | 1.63 | 0.61 |
| $C_2H_7NO_2$ | 2.73 | 0.43 |
| $C_5H_3N$ | 5.83 | 0.14 |
| | | |
| **78** | | |
| $CH_6N_2O_2$ | 2.02 | 0.41 |
| $C_2H_6O_3$ | 2.38 | 0.62 |
| $C_4H_2N_2$ | 5.12 | 0.11 |
| $C_5H_2O$ | 5.47 | 0.32 |
| $C_6H_6$ | 6.58 | 0.18 |
| | | |
| **79** | | |
| $CH_5NO_3$ | 1.66 | 0.61 |
| $C_5H_5N$ | 5.87 | 0.14 |
| | | |
| **80** | | |
| $C_4H_4N_2$ | 5.15 | 0.11 |
| $C_5H_4O$ | 5.51 | 0.32 |
| $C_6H_8$ | 6.61 | 0.18 |
| | | |
| **81** | | |
| $C_3H_3N_3$ | 4.43 | 0.08 |
| $C_4H_3NO$ | 4.79 | 0.29 |
| $C_5H_7N$ | 5.90 | 0.14 |
| | | |
| **82** | | |
| $C_3H_2N_2O$ | 4.08 | 0.36 |
| $C_4H_2O_2$ | 4.43 | 0.48 |
| $C_4H_6N_2$ | 4.18 | 0.11 |
| $C_5H_6O$ | 5.54 | 0.32 |
| $C_6H_{10}$ | 6.64 | 0.19 |
| | | |
| **83** | | |
| $C_3HNO_2$ | 3.72 | 0.45 |
| $C_3H_5N_3$ | 4.47 | 0.08 |
| $C_4H_5NO$ | 4.82 | 0.29 |
| $C_5H_9N$ | 5.93 | 0.15 |

| | $M+1$ | $M+2$ |
|---|---|---|
| **84** | | |
| $C_3H_4N_2O$ | 4.11 | 0.27 |
| $C_4H_4O_2$ | 4.47 | 0.48 |
| $C_4H_8N_2$ | 5.21 | 0.11 |
| $C_5H_8O$ | 5.57 | 0.33 |
| $C_6H_{12}$ | 6.68 | 0.19 |
| | | |
| **85** | | |
| $C_2H_3N_3O$ | 3.39 | 0.24 |
| $C_3H_3NO_2$ | 3.75 | 0.45 |
| $C_3H_7N_3$ | 4.50 | 0.08 |
| $C_4H_7NO$ | 4.86 | 0.29 |
| $C_5H_{11}N$ | 5.96 | 0.15 |
| | | |
| **86** | | |
| $C_2H_2N_2O_2$ | 3.03 | 0.43 |
| $C_3H_6N_2O$ | 4.14 | 0.27 |
| $C_4H_6O_2$ | 4.50 | 0.48 |
| $C_4H_{10}N_2$ | 5.25 | 0.11 |
| $C_5H_{10}O$ | 5.60 | 0.33 |
| $C_6H_{14}$ | 6.71 | 0.19 |
| | | |
| **87** | | |
| $C_2H_5N_3O$ | 3.43 | 0.25 |
| $C_3H_5NO_2$ | 3.78 | 0.45 |
| $C_3H_9N_3$ | 4.53 | 0.08 |
| $C_4H_9NO$ | 4.89 | 0.30 |
| $C_5H_{13}N$ | 5.99 | 0.15 |
| | | |
| **88** | | |
| $C_2H_4N_2O_2$ | 3.07 | 0.43 |
| $C_2H_8N_4$ | 3.82 | 0.06 |
| $C_3H_4O_3$ | 3.42 | 0.64 |
| $C_3H_8N_2O$ | 4.17 | 0.27 |
| $C_4H_8O_2$ | 4.53 | 0.48 |
| $C_4H_{12}N_2$ | 5.28 | 0.11 |
| $C_5H_{12}O$ | 5.63 | 0.33 |
| $C_7H_4$ | 7.63 | 0.25 |
| | | |
| **89** | | |
| $CH_3N_3O_2$ | 2.35 | 0.42 |
| $C_2H_3NO_3$ | 2.71 | 0.63 |
| $C_2H_7N_3O$ | 3.46 | 0.25 |
| $C_3H_7NO_2$ | 3.81 | 0.46 |
| $C_3H_{11}N_3$ | 4.56 | 0.84 |
| $C_4H_{11}NO$ | 4.92 | 0.30 |
| $C_6H_3N$ | 6.91 | 0.20 |

| 90 | $M+1$ | $M+2$ |
|---|---|---|
| $CH_2N_2O_3$ | 1.99 | 0.61 |
| $C_2H_2O_4$ | 2.35 | 0.82 |
| $C_2H_6N_2O_2$ | 3.10 | 0.44 |
| $C_3H_6O_3$ | 3.46 | 0.64 |
| $C_3H_{10}N_2O$ | 4.20 | 0.27 |
| $C_4H_{10}O_2$ | 4.56 | 0.48 |
| $C_5H_2N_2$ | 6.20 | 0.16 |
| $C_6H_2O$ | 6.56 | 0.38 |
| $C_7H_6$ | 7.66 | 0.25 |

| 91 | $M+1$ | $M+2$ |
|---|---|---|
| $CH_5N_3O_2$ | 2.38 | 0.42 |
| $C_2H_5NO_3$ | 2.74 | 0.63 |
| $C_2H_9N_3O$ | 3.49 | 0.25 |
| $C_3H_9NO_2$ | 3.85 | 0.46 |
| $C_6H_5N$ | 6.95 | 0.21 |

| 92 | $M+1$ | $M+2$ |
|---|---|---|
| $CH_4N_2O_3$ | 2.03 | 0.61 |
| $C_2H_4O_4$ | 2.38 | 0.82 |
| $C_2H_8N_2O_2$ | 3.13 | 0.44 |
| $C_3H_8O_3$ | 3.49 | 0.64 |
| $C_5H_4N_2$ | 6.23 | 0.16 |
| $C_6H_4O$ | 6.59 | 0.38 |
| $C_7H_8$ | 7.69 | 0.26 |

| 93 | $M+1$ | $M+2$ |
|---|---|---|
| $CH_7N_3O_2$ | 2.42 | 0.42 |
| $C_2H_7NO_3$ | 2.77 | 0.63 |
| $C_4H_3N_3$ | 5.52 | 0.13 |
| $C_5H_3NO$ | 5.87 | 0.34 |
| $C_6H_7N$ | 6.98 | 0.21 |

| 94 | $M+1$ | $M+2$ |
|---|---|---|
| $CH_6N_2O_3$ | 2.06 | 0.62 |
| $C_2H_6O_4$ | 2.41 | 0.82 |
| $C_3H_2N_4$ | 4.80 | 0.09 |
| $C_4H_2N_2O$ | 5.16 | 0.31 |
| $C_5H_2O_2$ | 5.51 | 0.52 |
| $C_5H_6N_2$ | 6.26 | 0.17 |
| $C_6H_6O$ | 6.62 | 0.38 |
| $C_7H_{10}$ | 7.72 | 0.26 |

| 95 | $M+1$ | $M+2$ |
|---|---|---|
| $C_4H_5N_3$ | 5.55 | 0.13 |
| $C_5H_5NO$ | 5.90 | 0.34 |
| $C_6H_9N$ | 7.01 | 0.21 |

| 96 | $M+1$ | $M+2$ |
|---|---|---|
| $C_3H_4N_4$ | 4.83 | 0.10 |
| $C_4H_4N_2O$ | 5.19 | 0.31 |
| $C_5H_4O_2$ | 5.55 | 0.53 |
| $C_5H_8N_2$ | 6.29 | 0.17 |
| $C_6H_8O$ | 6.65 | 0.39 |
| $C_7H_{12}$ | 7.76 | 0.26 |

| 97 | $M+1$ | $M+2$ |
|---|---|---|
| $C_3H_3N_3O$ | 4.47 | 0.28 |
| $C_4H_3NO_2$ | 4.83 | 0.49 |
| $C_4H_7N_3$ | 5.58 | 0.13 |
| $C_5H_7NO$ | 5.94 | 0.35 |
| $C_6H_{11}N$ | 7.04 | 0.21 |

| 98 | $M+1$ | $M+2$ |
|---|---|---|
| $C_3H_2N_2O_2$ | 4.12 | 0.47 |
| $C_3H_6N_4$ | 4.86 | 0.10 |
| $C_4H_2O_3$ | 4.47 | 0.68 |
| $C_4H_6N_2O$ | 5.22 | 0.31 |
| $C_5H_6O_2$ | 5.58 | 0.53 |
| $C_5H_{10}N_2$ | 6.33 | 0.17 |
| $C_6H_{10}O$ | 6.68 | 0.39 |
| $C_7H_{14}$ | 7.79 | 0.26 |

| 99 | $M+1$ | $M+2$ |
|---|---|---|
| $C_3H_5N_3O$ | 4.51 | 0.28 |
| $C_4H_5NO_2$ | 4.86 | 0.50 |
| $C_4H_9N_3$ | 5.61 | 0.13 |
| $C_5H_9NO$ | 5.97 | 0.35 |
| $C_6H_{13}N$ | 7.07 | 0.21 |

| 100 | $M+1$ | $M+2$ |
|---|---|---|
| $C_3H_4N_2O_2$ | 4.15 | 0.47 |
| $C_3H_8N_4$ | 4.90 | 0.10 |
| $C_4H_4O_3$ | 4.50 | 0.68 |
| $C_4H_8N_2O$ | 5.25 | 0.31 |
| $C_5H_8O_2$ | 5.61 | 0.53 |
| $C_5H_{12}N_2$ | 6.36 | 0.17 |
| $C_6H_{12}O$ | 6.72 | 0.39 |
| $C_7H_{16}$ | 7.82 | 0.26 |
| $C_8H_4$ | 8.71 | 0.33 |

| 101 | $M+1$ | $M+2$ |
|---|---|---|
| $C_2H_3N_3O_2$ | 3.43 | 0.45 |
| $C_3H_3NO_3$ | 3.79 | 0.65 |
| $C_3H_7N_3O$ | 4.54 | 0.28 |

| 102 | $M+1$ | $M+2$ |
|---|---|---|
| $C_4H_7NO_2$ | 4.89 | 0.50 |
| $C_4H_{11}N_3$ | 5.64 | 0.13 |
| $C_5H_{11}NO$ | 6.00 | 0.35 |
| $C_6H_{15}N$ | 7.11 | 0.22 |

| 102 | $M+1$ | $M+2$ |
|---|---|---|
| $C_2H_2N_2O_3$ | 3.07 | 0.64 |
| $C_2H_6N_4O$ | 3.82 | 0.26 |
| $C_3H_2O_4$ | 3.43 | 0.84 |
| $C_3H_6N_2O_2$ | 4.18 | 0.47 |
| $C_3H_{10}N_4$ | 4.93 | 0.10 |
| $C_4H_6O_3$ | 4.54 | 0.68 |
| $C_4H_{10}N_2O$ | 5.28 | 0.32 |
| $C_5H_{10}O_2$ | 5.64 | 0.53 |
| $C_5H_{14}N_2$ | 6.39 | 0.17 |
| $C_6H_{14}O$ | 6.75 | 0.39 |
| $C_8H_6$ | 8.74 | 0.34 |

| 103 | $M+1$ | $M+2$ |
|---|---|---|
| $C_2H_5N_3O_2$ | 3.46 | 0.45 |
| $C_3H_5NO_3$ | 3.82 | 0.66 |
| $C_3H_9N_3O$ | 4.57 | 0.29 |
| $C_4H_9NO_2$ | 4.93 | 0.50 |
| $C_4H_{13}N_3$ | 5.68 | 0.14 |
| $C_5H_{13}NO$ | 6.03 | 0.35 |
| $C_7H_5N$ | 8.03 | 0.28 |

| 104 | $M+1$ | $M+2$ |
|---|---|---|
| $C_2H_4N_2O_3$ | 3.11 | 0.64 |
| $C_2H_8N_4O$ | 3.85 | 0.26 |
| $C_3H_4O_4$ | 3.46 | 0.84 |
| $C_3H_8N_2O_2$ | 4.21 | 0.47 |
| $C_3H_{12}N_4$ | 4.96 | 0.10 |
| $C_4H_{12}N_2O$ | 5.32 | 0.32 |
| $C_5H_{12}O_2$ | 5.67 | 0.53 |
| $C_6H_4N_2$ | 7.31 | 0.23 |
| $C_7H_4O$ | 7.67 | 0.45 |
| $C_8H_8$ | 8.77 | 0.34 |

| 105 | $M+1$ | $M+2$ |
|---|---|---|
| $C_2H_3NO_4$ | 2.75 | 0.83 |
| $C_2H_7N_3O_2$ | 3.50 | 0.45 |
| $C_3H_7NO_3$ | 3.85 | 0.66 |
| $C_3H_{11}N_3O$ | 4.60 | 0.29 |
| $C_4H_{11}NO_2$ | 4.96 | 0.50 |
| $C_5H_3N_3$ | 6.60 | 0.19 |
| $C_6H_3NO$ | 6.95 | 0.41 |
| $C_7H_7N$ | 8.06 | 0.28 |

| | M + 1 | M + 2 |
|---|---|---|
| **106** | | |
| $CH_2N_2O_4$ | 2.03 | 0.82 |
| $C_2H_6N_2O_3$ | 3.14 | 0.64 |
| $C_3H_6O_4$ | 3.49 | 0.85 |
| $C_3H_{10}N_2O_2$ | 4.24 | 0.47 |
| $C_4H_{10}O_3$ | 4.60 | 0.68 |
| $C_5H_2N_2O$ | 6.24 | 0.36 |
| $C_6H_2O_2$ | 6.59 | 0.58 |
| $C_6H_6N_2$ | 7.34 | 0.23 |
| $C_7H_6O$ | 7.70 | 0.46 |
| $C_8H_{10}$ | 8.81 | 0.34 |
| | | |
| **107** | | |
| $C_2H_9N_3O_2$ | 3.53 | 0.45 |
| $C_3H_8NO_3$ | 3.89 | 0.66 |
| $C_5H_5N_3$ | 6.63 | 0.19 |
| $C_6H_5NO$ | 6.98 | 0.41 |
| $C_7H_9N$ | 8.09 | 0.29 |
| | | |
| **108** | | |
| $C_2H_8N_2O_3$ | 3.17 | 0.64 |
| $C_3H_8O_4$ | 3.53 | 0.85 |
| $C_5H_4N_2O$ | 6.27 | 0.37 |
| $C_6H_4O_2$ | 6.63 | 0.59 |
| $C_6H_8N_2$ | 7.38 | 0.24 |
| $C_7H_8O$ | 7.73 | 0.46 |
| $C_8H_{12}$ | 8.84 | 0.34 |
| | | |
| **109** | | |
| $CH_7N_3O_3$ | 2.45 | 0.62 |
| $C_2H_7NO_4$ | 2.81 | 0.83 |
| $C_4H_3N_3O$ | 5.55 | 0.33 |
| $C_5H_3NO_2$ | 5.91 | 0.55 |
| $C_5H_7N_3$ | 6.66 | 0.19 |
| $C_6H_7NO$ | 7.02 | 0.41 |
| $C_7H_{11}N$ | 8.12 | 0.29 |
| | | |
| **110** | | |
| $CH_6N_2O_4$ | 2.10 | 0.82 |
| $C_5H_6N_2O$ | 6.30 | 0.37 |
| $C_6H_6O_2$ | 6.66 | 0.59 |
| $C_6H_{10}N_2$ | 7.41 | 0.24 |
| $C_7H_{10}O$ | 7.76 | 0.46 |
| $C_9H_{14}$ | 8.87 | 0.35 |
| | | |
| **111** | | |
| $C_5H_5NO_2$ | 5.94 | 0.55 |
| $C_5H_9N_3$ | 6.69 | 0.19 |
| $C_6H_9NO$ | 7.05 | 0.41 |
| $C_7H_{13}N$ | 8.15 | 0.29 |

| | M + 1 | M + 2 |
|---|---|---|
| **112** | | |
| $C_4H_4N_2O_2$ | 5.23 | 0.51 |
| $C_5H_4O_3$ | 5.58 | 0.73 |
| $C_5H_8N_2O$ | 6.33 | 0.37 |
| $C_6H_8O_2$ | 6.69 | 0.59 |
| $C_6H_{12}N_2$ | 7.44 | 0.24 |
| $C_7H_{12}O$ | 7.80 | 0.46 |
| $C_8H_{16}$ | 8.90 | 0.35 |
| $C_9H_4$ | 9.79 | 0.43 |
| | | |
| **113** | | |
| $C_3H_3N_3O_2$ | 4.51 | 0.48 |
| $C_4H_3NO_3$ | 4.87 | 0.70 |
| $C_4H_7N_3O$ | 5.62 | 0.33 |
| $C_5H_7NO_2$ | 5.98 | 0.55 |
| $C_5H_{11}N_3$ | 6.72 | 0.19 |
| $C_6H_{11}NO$ | 7.08 | 0.42 |
| $C_7H_{15}N$ | 8.19 | 0.29 |
| $C_8H_3N$ | 9.07 | 0.36 |
| | | |
| **114** | | |
| $C_3H_2N_2O_3$ | 4.15 | 0.67 |
| $C_4H_6N_2O_2$ | 5.26 | 0.51 |
| $C_4H_{10}N_4$ | 6.01 | 0.15 |
| $C_5H_6O_3$ | 5.62 | 0.73 |
| $C_5H_{10}N_2O$ | 6.37 | 0.37 |
| $C_6H_{10}O_2$ | 6.72 | 0.59 |
| $C_6H_{14}N_2$ | 7.47 | 0.24 |
| $C_7H_2N_2$ | 8.36 | 0.31 |
| $C_7H_{14}O$ | 7.83 | 0.47 |
| $C_8H_{18}$ | 8.93 | 0.35 |
| $C_9H_6$ | 9.82 | 0.43 |
| | | |
| **115** | | |
| $C_3H_5N_3O_2$ | 4.54 | 0.48 |
| $C_4H_5NO_3$ | 4.90 | 0.70 |
| $C_4H_9N_3O$ | 5.65 | 0.33 |
| $C_5H_9NO_2$ | 6.01 | 0.55 |
| $C_5H_{13}N_3$ | 6.76 | 0.20 |
| $C_6H_{13}NO$ | 7.11 | 0.42 |
| $C_7H_{17}N$ | 8.22 | 0.30 |
| $C_8H_5N$ | 9.11 | 0.37 |
| | | |
| **116** | | |
| $C_3H_4N_2O_3$ | 4.19 | 0.67 |
| $C_3H_8N_4O$ | 4.94 | 0.30 |
| $C_4H_4O_4$ | 4.54 | 0.88 |
| $C_4H_8N_2O_2$ | 5.29 | 0.52 |
| $C_4H_{12}N_4$ | 6.04 | 0.16 |

| | M + 1 | M + 2 |
|---|---|---|
| $C_5H_8O_3$ | 5.65 | 0.73 |
| $C_5H_{12}N_2O$ | 6.40 | 0.37 |
| $C_6H_{12}O_2$ | 6.75 | 0.59 |
| $C_6H_{16}N_2$ | 7.50 | 0.24 |
| $C_7H_4N_2$ | 8.39 | 0.31 |
| $C_7H_{16}O$ | 7.86 | 0.47 |
| $C_8H_4O$ | 8.75 | 0.54 |
| $C_9H_8$ | 9.85 | 0.43 |
| | | |
| **117** | | |
| $C_3H_3NO_4$ | 3.83 | 0.86 |
| $C_3H_7N_3O_2$ | 4.58 | 0.49 |
| $C_4H_7NO_3$ | 4.93 | 0.70 |
| $C_4H_{11}N_3O$ | 5.68 | 0.34 |
| $C_5H_{11}NO_2$ | 6.04 | 0.55 |
| $C_5H_{15}N_3$ | 6.79 | 0.20 |
| $C_6H_3N_3$ | 7.68 | 0.26 |
| $C_6H_{15}NO$ | 7.14 | 0.42 |
| $C_7H_3NO$ | 8.03 | 0.48 |
| $C_8H_7N$ | 9.14 | 0.37 |
| | | |
| **118** | | |
| $C_3H_6N_2O_3$ | 4.22 | 0.67 |
| $C_3H_{10}N_4O$ | 4.97 | 0.30 |
| $C_4H_6O_4$ | 4.58 | 0.88 |
| $C_4H_{10}N_2O_2$ | 5.32 | 0.52 |
| $C_4H_{14}N_4$ | 6.07 | 0.16 |
| $C_5H_{10}O_3$ | 5.68 | 0.73 |
| $C_5H_{14}N_2O$ | 6.43 | 0.38 |
| $C_6H_{14}O_2$ | 6.79 | 0.60 |
| $C_7H_6N_2$ | 8.42 | 0.31 |
| $C_9H_6O$ | 8.78 | 0.54 |
| $C_9H_{10}$ | 9.89 | 0.44 |
| | | |
| **119** | | |
| $C_3H_5NO_4$ | 3.86 | 0.86 |
| $C_3H_9N_3O_2$ | 4.61 | 0.49 |
| $C_4H_9NO_3$ | 4.97 | 0.70 |
| $C_4H_{13}N_3O$ | 5.71 | 0.34 |
| $C_5H_{13}NO_2$ | 6.07 | 0.56 |
| $C_6H_5N_3$ | 7.71 | 0.26 |
| $C_7H_5NO$ | 8.07 | 0.48 |
| $C_8H_9N$ | 9.17 | 0.37 |
| | | |
| **120** | | |
| $C_2H_4N_2O_4$ | 3.15 | 0.84 |
| $C_3H_8N_2O_3$ | 4.25 | 0.67 |
| $C_3H_{12}N_4O$ | 5.00 | 0.31 |
| $C_4H_8O_4$ | 4.61 | 0.88 |
| $C_4H_{12}N_2O_2$ | 5.36 | 0.52 |

|  | M + 1 | M + 2 |
|---|---|---|
| $C_5H_{12}O_3$ | 5.71 | 0.74 |
| $C_6H_4N_2O$ | 7.35 | 0.43 |
| $C_7H_4O_2$ | 7.71 | 0.66 |
| $C_7H_8N_2$ | 8.46 | 0.32 |
| $C_8H_8O$ | 8.81 | 0.54 |
| $C_9H_{12}$ | 9.92 | 0.44 |
| **121** | | |
| $C_3H_7NO_4$ | 3.89 | 0.86 |
| $C_3H_{11}N_3O_2$ | 4.64 | 0.49 |
| $C_4H_{11}NO_3$ | 5.00 | 0.70 |
| $C_5H_3N_3O$ | 6.64 | 0.39 |
| $C_6H_3NO_2$ | 6.99 | 0.61 |
| $C_6H_7N_3$ | 7.74 | 0.26 |
| $C_7H_7NO$ | 8.10 | 0.49 |
| $C_8H_{11}N$ | 9.20 | 0.38 |
| **122** | | |
| $C_2H_6N_2O_4$ | 3.18 | 0.84 |
| $C_2H_{10}N_4O_2$ | 3.93 | 0.46 |
| $C_3H_{10}N_2O_3$ | 4.28 | 0.67 |
| $C_4H_{10}O_4$ | 4.64 | 0.89 |
| $C_5H_6N_4$ | 7.03 | 0.21 |
| $C_6H_2O_3$ | 6.63 | 0.79 |
| $C_6H_6N_2O$ | 7.38 | 0.44 |
| $C_7H_6O_2$ | 7.74 | 0.66 |
| $C_7H_{10}N_2$ | 8.49 | 0.32 |
| $C_8H_{10}O$ | 8.84 | 0.54 |
| $C_9H_{14}$ | 9.95 | 0.44 |
| **123** | | |
| $C_3H_9NO_4$ | 3.92 | 0.86 |
| $C_5H_5N_3O$ | 6.67 | 0.39 |
| $C_6H_5NO_2$ | 7.02 | 0.61 |
| $C_6H_9N_3$ | 7.77 | 0.26 |
| $C_7H_9NO$ | 8.13 | 0.49 |
| $C_8H_{13}N$ | 9.23 | 0.38 |
| **124** | | |
| $C_4H_4N_4O$ | 5.95 | 0.35 |
| $C_5H_4N_2O_2$ | 6.31 | 0.57 |
| $C_5H_8N_4$ | 7.06 | 0.22 |
| $C_6H_4O_3$ | 6.67 | 0.79 |
| $C_6H_8N_2O$ | 7.41 | 0.44 |
| $C_7H_8O_2$ | 7.77 | 0.66 |
| $C_7H_{12}N_2$ | 8.52 | 0.32 |
| $C_8H_{12}O$ | 8.88 | 0.55 |
| $C_9H_{16}$ | 9.98 | 0.45 |
| $C_{10}H_4$ | 10.87 | 0.53 |

|  | M + 1 | M + 2 |
|---|---|---|
| **125** | | |
| $C_4H_3N_3O_2$ | 5.59 | 0.53 |
| $C_5H_3NO_3$ | 5.95 | 0.75 |
| $C_5H_7N_3O$ | 6.70 | 0.39 |
| $C_6H_7NO_2$ | 7.06 | 0.61 |
| $C_6H_{11}N_3$ | 7.80 | 0.27 |
| $C_7H_{11}NO$ | 8.16 | 0.49 |
| $C_8H_{15}N$ | 9.27 | 0.38 |
| $C_9H_3N$ | 10.16 | 0.46 |
| **126** | | |
| $C_4H_2N_2O_3$ | 5.24 | 0.71 |
| $C_4H_6N_4O$ | 5.98 | 0.35 |
| $C_5H_6N_2O_2$ | 6.34 | 0.57 |
| $C_5H_{10}N_4$ | 7.09 | 0.22 |
| $C_6H_6O_3$ | 6.70 | 0.79 |
| $C_6H_{10}N_2O$ | 7.45 | 0.44 |
| $C_7H_{10}O_2$ | 7.80 | 0.66 |
| $C_7H_{14}N_2$ | 8.55 | 0.32 |
| $C_8H_2N_2$ | 9.44 | 0.40 |
| $C_8H_{14}O$ | 8.91 | 0.55 |
| $C_9H_{18}$ | 10.01 | 0.45 |
| $C_{10}H_6$ | 10.90 | 0.54 |
| **127** | | |
| $C_4H_5N_3O_2$ | 5.63 | 0.53 |
| $C_5H_5NO_3$ | 5.98 | 0.75 |
| $C_5H_9N_3O$ | 6.73 | 0.40 |
| $C_6H_9NO_2$ | 7.09 | 0.62 |
| $C_6H_{13}N_3$ | 7.84 | 0.27 |
| $C_7H_{13}NO$ | 8.19 | 0.49 |
| $C_8H_{17}N$ | 9.30 | 0.38 |
| $C_9H_5N$ | 10.19 | 0.47 |
| **128** | | |
| $C_3H_4N_4O_2$ | 4.91 | 0.50 |
| $C_4H_4N_2O_3$ | 5.27 | 0.72 |
| $C_4H_8N_4O$ | 6.02 | 0.36 |
| $C_5H_8N_2O_2$ | 6.37 | 0.57 |
| $C_5H_{12}N_4$ | 7.12 | 0.22 |
| $C_6H_8O_3$ | 6.73 | 0.79 |
| $C_6H_{12}N_2O$ | 7.48 | 0.44 |
| $C_7H_{12}O_2$ | 7.83 | 0.67 |
| $C_7H_{16}N_2$ | 8.58 | 0.33 |
| $C_8H_4N_2$ | 9.47 | 0.40 |
| $C_8H_{16}O$ | 8.94 | 0.55 |
| $C_9H_4O$ | 9.83 | 0.63 |
| $C_9H_{20}$ | 10.05 | 0.45 |
| $C_{10}H_8$ | 10.94 | 0.54 |

|  | M + 1 | M + 2 |
|---|---|---|
| **129** | | |
| $C_4H_7N_3O_2$ | 5.66 | 0.54 |
| $C_5H_7NO_3$ | 6.01 | 0.75 |
| $C_5H_{11}N_3O$ | 6.76 | 0.40 |
| $C_6H_{11}NO_2$ | 7.12 | 0.62 |
| $C_6H_{15}N_3$ | 7.87 | 0.27 |
| $C_7H_{15}NO$ | 8.23 | 0.50 |
| $C_8H_3NO$ | 9.11 | 0.57 |
| $C_8H_{19}N$ | 9.33 | 0.39 |
| $C_9H_7N$ | 10.22 | 0.47 |
| **130** | | |
| $C_4H_6N_2O_3$ | 5.30 | 0.72 |
| $C_4H_{10}N_4O$ | 6.05 | 0.36 |
| $C_5H_6O_4$ | 5.66 | 0.93 |
| $C_5H_{10}N_2O_2$ | 6.40 | 0.58 |
| $C_5H_{14}N_4$ | 7.15 | 0.22 |
| $C_6H_{10}O_3$ | 6.76 | 0.79 |
| $C_6H_{14}N_2O$ | 7.51 | 0.45 |
| $C_7H_{14}O_2$ | 7.87 | 0.67 |
| $C_7H_{18}N_2$ | 8.62 | 0.33 |
| $C_8H_6N_2$ | 9.50 | 0.40 |
| $C_8H_{18}O$ | 8.97 | 0.56 |
| $C_9H_6O$ | 9.86 | 0.63 |
| $C_{10}H_{10}$ | 10.97 | 0.54 |
| **131** | | |
| $C_4H_5NO_4$ | 4.94 | 0.90 |
| $C_4H_9N_3O_2$ | 5.69 | 0.54 |
| $C_5H_9NO_3$ | 6.05 | 0.75 |
| $C_5H_{13}N_3O$ | 6.80 | 0.40 |
| $C_6H_{13}NO_2$ | 7.15 | 0.62 |
| $C_6H_{17}N_3$ | 7.90 | 0.27 |
| $C_7H_5N_3$ | 8.79 | 0.34 |
| $C_7H_{17}NO$ | 8.26 | 0.50 |
| $C_8H_5NO$ | 9.15 | 0.57 |
| $C_9H_9N$ | 10.25 | 0.47 |
| **132** | | |
| $C_3H_4N_2O_4$ | 4.23 | 0.87 |
| $C_3H_8N_4O_2$ | 4.97 | 0.50 |
| $C_4H_8N_2O_3$ | 5.33 | 0.72 |
| $C_4H_{12}N_4O$ | 6.08 | 0.36 |
| $C_5H_8O_4$ | 5.69 | 0.93 |
| $C_5H_{12}N_2O_2$ | 6.44 | 0.58 |
| $C_5H_{16}N_4$ | 7.19 | 0.23 |
| $C_6H_4N_4$ | 8.07 | 0.29 |
| $C_6H_{12}O_3$ | 6.97 | 0.80 |
| $C_6H_{16}N_2O$ | 7.54 | 0.45 |
| $C_7H_4N_2O$ | 8.43 | 0.51 |

| | M + 1 | M + 2 |
|---|---|---|
| $C_7H_{16}O_2$ | 7.90 | 0.67 |
| $C_8H_4O_2$ | 8.79 | 0.74 |
| $C_8H_8N_2$ | 9.54 | 0.41 |
| $C_9H_8O$ | 9.89 | 0.64 |
| $C_{10}H_{12}$ | 11.00 | 0.55 |

**133**

| | M + 1 | M + 2 |
|---|---|---|
| $C_4H_7NO_4$ | 4.97 | 0.90 |
| $C_4H_{11}N_3O_2$ | 5.72 | 0.54 |
| $C_5H_{11}NO_3$ | 6.08 | 0.76 |
| $C_5H_{15}N_3O$ | 6.83 | 0.40 |
| $C_6H_3N_3O$ | 7.72 | 0.46 |
| $C_6H_{15}NO_2$ | 7.18 | 0.62 |
| $C_7H_3NO_2$ | 8.07 | 0.69 |
| $C_7H_7N_3$ | 8.82 | 0.35 |
| $C_8H_7NO$ | 9.18 | 0.57 |
| $C_9H_{11}N$ | 10.28 | 0.48 |

**134**

| | M + 1 | M + 2 |
|---|---|---|
| $C_4H_{10}N_2O_3$ | 5.36 | 0.72 |
| $C_4H_{14}N_4O$ | 6.11 | 0.36 |
| $C_5H_{10}O_4$ | 5.72 | 0.94 |
| $C_5H_{14}N_2O_2$ | 6.47 | 0.58 |
| $C_6H_6N_4$ | 8.11 | 0.29 |
| $C_6H_{14}O_3$ | 6.83 | 0.80 |
| $C_7H_6N_2O$ | 8.46 | 0.52 |
| $C_8H_4O_2$ | 8.82 | 0.74 |
| $C_8H_{10}N_2$ | 9.57 | 0.41 |
| $C_9H_{10}O$ | 9.93 | 0.64 |
| $C_{10}H_{14}$ | 11.03 | 0.55 |

**135**

| | M + 1 | M + 2 |
|---|---|---|
| $C_4H_9NO_4$ | 5.00 | 0.90 |
| $C_4H_{13}N_3O_2$ | 5.75 | 0.54 |
| $C_5H_{13}NO_3$ | 6.11 | 0.76 |
| $C_6H_5N_3O$ | 7.75 | 0.46 |
| $C_7H_5NO_2$ | 8.10 | 0.69 |
| $C_7H_9N_3$ | 8.85 | 0.35 |
| $C_8H_9NO$ | 9.21 | 0.58 |
| $C_9H_{13}N$ | 10.32 | 0.48 |

**136**

| | M + 1 | M + 2 |
|---|---|---|
| $C_3H_8N_2O_4$ | 4.29 | 0.87 |
| $C_3H_{12}N_4O_2$ | 5.04 | 0.51 |
| $C_4H_{12}N_2O_3$ | 5.40 | 0.72 |
| $C_5H_{12}O_4$ | 5.75 | 0.94 |
| $C_6H_4N_2O_2$ | 7.39 | 0.64 |
| $C_6H_8N_4$ | 8.14 | 0.29 |

| | M + 1 | M + 2 |
|---|---|---|
| $C_7H_4O_3$ | 7.75 | 0.86 |
| $C_7H_8N_2O$ | 8.49 | 0.52 |
| $C_8H_8O_2$ | 8.85 | 0.75 |
| $C_8H_{12}N_2$ | 9.60 | 0.41 |
| $C_9H_{12}O$ | 9.96 | 0.64 |
| $C_{10}H_{16}$ | 11.06 | 0.55 |

**137**

| | M + 1 | M + 2 |
|---|---|---|
| $C_3H_{11}N_3O_3$ | 4.68 | 0.69 |
| $C_4H_{11}NO_4$ | 5.04 | 0.90 |
| $C_6H_3NO_3$ | 7.03 | 0.81 |
| $C_6H_7N_3O$ | 7.78 | 0.47 |
| $C_7H_7NO_2$ | 8.14 | 0.69 |
| $C_7H_{11}N_3$ | 8.89 | 0.35 |
| $C_8H_{11}NO$ | 9.24 | 0.58 |
| $C_9H_{15}N$ | 10.35 | 0.48 |

**138**

| | M + 1 | M + 2 |
|---|---|---|
| $C_3H_{10}N_2O_4$ | 4.32 | 0.88 |
| $C_5H_6N_4O$ | 7.06 | 0.42 |
| $C_6H_6N_2O_2$ | 7.42 | 0.64 |
| $C_6H_{10}N_4$ | 8.17 | 0.30 |
| $C_7H_6O_3$ | 7.78 | 0.86 |
| $C_7H_{10}N_2O$ | 8.53 | 0.52 |
| $C_8H_{10}O_2$ | 8.88 | 0.75 |
| $C_8H_{14}N_2$ | 9.63 | 0.42 |
| $C_9H_{14}O$ | 9.99 | 0.65 |
| $C_{10}H_{18}$ | 11.09 | 0.56 |
| $C_{11}H_6$ | 11.98 | 0.65 |

**139**

| | M + 1 | M + 2 |
|---|---|---|
| $C_4H_3N_3O_3$ | 5.60 | 0.73 |
| $C_5H_5N_3O_2$ | 6.71 | 0.59 |
| $C_6H_5NO_3$ | 7.06 | 0.82 |
| $C_6H_9N_3O$ | 7.81 | 0.47 |
| $C_7H_9NO_2$ | 8.17 | 0.69 |
| $C_7H_{13}N_3$ | 8.92 | 0.35 |
| $C_8H_{13}NO$ | 9.27 | 0.58 |
| $C_9H_{17}N$ | 10.38 | 0.49 |
| $C_{10}H_5N$ | 11.27 | 0.58 |

**140**

| | M + 1 | M + 2 |
|---|---|---|
| $C_5H_4N_2O_3$ | 6.35 | 0.77 |
| $C_5H_8N_4O$ | 7.10 | 0.42 |
| $C_6H_4O_4$ | 6.70 | 0.99 |
| $C_6H_8N_2O_2$ | 7.45 | 0.64 |
| $C_6H_{12}N_4$ | 8.20 | 0.30 |
| $C_7H_8O_3$ | 7.81 | 0.87 |

| | M + 1 | M + 2 |
|---|---|---|
| $C_7H_{12}N_2O$ | 8.56 | 0.52 |
| $C_8H_{12}O_2$ | 8.92 | 0.75 |
| $C_8H_{16}N_2$ | 9.66 | 0.42 |
| $C_9H_4N_2$ | 10.55 | 0.50 |
| $C_9H_{16}O$ | 10.02 | 0.65 |
| $C_{10}H_4O$ | 10.91 | 0.74 |
| $C_{10}H_{20}$ | 11.13 | 0.56 |
| $C_{11}H_8$ | 12.02 | 0.66 |

**141**

| | M + 1 | M + 2 |
|---|---|---|
| $C_5H_3NO_4$ | 5.99 | 0.95 |
| $C_5H_7N_3O_2$ | 6.74 | 0.60 |
| $C_6H_7NO_3$ | 7.09 | 0.82 |
| $C_6H_{11}N_3O$ | 7.84 | 0.47 |
| $C_7H_{11}NO_2$ | 8.20 | 0.70 |
| $C_7H_{15}N_3$ | 8.95 | 0.36 |
| $C_8H_{15}NO$ | 9.31 | 0.59 |
| $C_9H_3NO$ | 10.19 | 0.67 |
| $C_9H_{19}N$ | 10.41 | 0.49 |
| $C_{10}H_7N$ | 11.30 | 0.58 |

**142**

| | M + 1 | M + 2 |
|---|---|---|
| $C_5H_6N_2O_3$ | 6.38 | 0.77 |
| $C_5H_{10}N_4O$ | 7.13 | 0.42 |
| $C_6H_6O_4$ | 6.74 | 0.99 |
| $C_6H_{10}N_2O_2$ | 7.49 | 0.64 |
| $C_6H_{14}N_4$ | 8.23 | 0.30 |
| $C_7H_{10}O_3$ | 7.84 | 0.87 |
| $C_7H_{14}N_2O$ | 8.59 | 0.53 |
| $C_8H_{14}O_2$ | 8.95 | 0.75 |
| $C_8H_{18}N_2$ | 9.70 | 0.42 |
| $C_9H_6N_2$ | 10.58 | 0.51 |
| $C_9H_{18}O$ | 10.05 | 0.65 |
| $C_{10}H_6O$ | 10.94 | 0.74 |
| $C_{10}H_{22}$ | 11.16 | 0.56 |
| $C_{11}H_{10}$ | 12.05 | 0.66 |

**143**

| | M + 1 | M + 2 |
|---|---|---|
| $C_5H_5NO_4$ | 6.02 | 0.95 |
| $C_5H_9N_3O_2$ | 6.77 | 0.60 |
| $C_6H_9NO_3$ | 7.13 | 0.82 |
| $C_6H_{13}N_3O$ | 7.88 | 0.47 |
| $C_7H_{13}NO_2$ | 8.23 | 0.70 |
| $C_7H_{17}N_3$ | 8.98 | 0.36 |
| $C_8H_5N_3$ | 9.87 | 0.44 |
| $C_8H_{17}NO$ | 9.34 | 0.59 |
| $C_9H_5NO$ | 10.23 | 0.67 |
| $C_9H_{21}N$ | 10.44 | 0.49 |
| $C_{10}H_9N$ | 11.33 | 0.58 |

| | $M+1$ | $M+2$ |
|---|---|---|
| **144** | | |
| $C_4H_8N_4O_2$ | 6.05 | 0.56 |
| $C_5H_8N_2O_3$ | 6.41 | 0.78 |
| $C_5H_{12}N_4O$ | 7.16 | 0.42 |
| $C_6H_8O_4$ | 6.77 | 1.00 |
| $C_6H_{12}N_2O_2$ | 7.52 | 0.65 |
| $C_6H_{16}N_4$ | 8.27 | 0.30 |
| $C_7H_{12}O_3$ | 7.87 | 0.87 |
| $C_7H_{16}N_2O$ | 8.62 | 0.53 |
| $C_8H_4N_2O$ | 9.51 | 0.60 |
| $C_8H_{16}O_2$ | 8.98 | 0.76 |
| $C_8H_{20}N_2$ | 9.73 | 0.43 |
| $C_9H_8N_2$ | 10.62 | 0.51 |
| $C_9H_{20}O$ | 10.09 | 0.66 |
| $C_{10}H_8O$ | 10.97 | 0.74 |
| $C_{11}H_{12}$ | 12.08 | 0.67 |
| | | |
| **145** | | |
| $C_4H_7N_3O_3$ | 5.70 | 0.74 |
| $C_5H_7NO_4$ | 6.05 | 0.96 |
| $C_5H_{11}N_3O_2$ | 6.80 | 0.60 |
| $C_6H_{11}NO_3$ | 7.16 | 0.82 |
| $C_6H_{15}N_3O$ | 7.91 | 0.48 |
| $C_7H_{13}O_3$ | 7.89 | 0.87 |
| $C_7H_{15}NO_2$ | 8.26 | 0.70 |
| $C_7H_{19}N_3$ | 9.01 | 0.36 |
| $C_8H_7N_3$ | 9.90 | 0.44 |
| $C_8H_{19}NO$ | 9.37 | 0.59 |
| $C_9H_7NO$ | 10.26 | 0.67 |
| $C_{10}H_{11}N$ | 11.36 | 0.59 |
| | | |
| **146** | | |
| $C_4H_6N_2O_4$ | 5.34 | 0.92 |
| $C_4H_{10}N_4O_2$ | 6.09 | 0.56 |
| $C_5H_{10}N_2O_3$ | 6.44 | 0.78 |
| $C_5H_{14}N_4O$ | 7.19 | 0.43 |
| $C_6H_{10}O_4$ | 6.80 | 1.00 |
| $C_6H_{14}N_2O_2$ | 7.55 | 0.65 |
| $C_6H_{18}N_4$ | 8.30 | 0.31 |
| $C_7H_6N_4$ | 9.19 | 0.38 |
| $C_7H_{14}O_3$ | 7.91 | 0.87 |
| $C_7H_{18}N_2O$ | 8.65 | 0.53 |
| $C_8N_6N_2O$ | 9.54 | 0.61 |
| $C_8H_{18}O_2$ | 9.01 | 0.76 |
| $C_9H_6O_2$ | 9.90 | 0.84 |
| $C_9H_{10}N_2$ | 10.65 | 0.51 |
| $C_{10}H_{10}O$ | 11.01 | 0.75 |
| $C_{11}H_{14}$ | 12.11 | 0.67 |

| | $M+1$ | $M+2$ |
|---|---|---|
| **147** | | |
| $C_4H_9N_3O_3$ | 5.73 | 0.74 |
| $C_5H_9NO_4$ | 6.09 | 0.96 |
| $C_5H_{13}N_3O_2$ | 6.83 | 0.60 |
| $C_6H_{13}NO_3$ | 7.19 | 0.82 |
| $C_6H_{17}N_3O$ | 7.94 | 0.48 |
| $C_7H_5N_3O$ | 8.83 | 0.55 |
| $C_7H_{17}NO_2$ | 8.30 | 0.70 |
| $C_8H_5NO_2$ | 9.19 | 0.78 |
| $C_8H_9N_3$ | 9.93 | 0.44 |
| $C_9H_9NO$ | 10.29 | 0.68 |
| $C_{10}H_{13}N$ | 11.40 | 0.59 |
| | | |
| **148** | | |
| $C_4H_8N_2O_4$ | 5.37 | 0.92 |
| $C_4H_{12}N_4O_2$ | 6.12 | 0.56 |
| $C_5H_{12}N_2O_3$ | 6.48 | 0.78 |
| $C_5H_{16}N_4O$ | 7.22 | 0.43 |
| $C_6H_4N_4O$ | 8.11 | 0.49 |
| $C_6H_{12}O_4$ | 6.83 | 1.00 |
| $C_6H_{16}N_2O_2$ | 7.58 | 0.65 |
| $C_7H_4N_2O_2$ | 8.47 | 0.72 |
| $C_7H_8N_4$ | 9.22 | 0.38 |
| $C_7H_{16}O_3$ | 7.94 | 0.88 |
| $C_8H_4O_3$ | 8.83 | 0.94 |
| $C_8H_8N_2O$ | 9.58 | 0.61 |
| $C_9H_8O_2$ | 9.93 | 0.84 |
| $C_9H_{12}N_2$ | 10.68 | 0.52 |
| $C_{10}H_{12}O$ | 11.04 | 0.75 |
| $C_{11}H_{16}$ | 12.14 | 0.67 |
| | | |
| **149** | | |
| $C_4H_{11}N_3O_3$ | 5.76 | 0.74 |
| $C_5H_{11}NO_4$ | 6.12 | 0.96 |
| $C_5H_{15}N_3O_2$ | 6.87 | 0.61 |
| $C_6H_{15}NO_3$ | 7.22 | 0.83 |
| $C_7H_3NO_3$ | 8.11 | 0.89 |
| $C_7H_7N_3O$ | 8.86 | 0.55 |
| $C_8H_7NO_2$ | 9.22 | 0.78 |
| $C_8H_{11}N_3$ | 9.97 | 0.45 |
| $C_9H_{11}NO$ | 10.32 | 0.68 |
| $C_{10}H_{15}N$ | 11.43 | 0.59 |
| | | |
| **150** | | |
| $C_4H_{10}N_2O_4$ | 5.40 | 0.92 |
| $C_4H_{14}N_4O_2$ | 6.15 | 0.56 |
| $C_5H_{14}N_2O_3$ | 6.51 | 0.78 |
| $C_6H_6N_4O$ | 8.15 | 0.49 |
| $C_6H_{14}O_4$ | 6.86 | 1.00 |

| | $M+1$ | $M+2$ |
|---|---|---|
| $C_7H_6N_2O_2$ | 8.50 | 0.72 |
| $C_7H_{10}N_4$ | 9.25 | 0.38 |
| $C_8H_6O_3$ | 8.86 | 0.95 |
| $C_8H_{10}N_2O$ | 9.61 | 0.61 |
| $C_9H_{10}O_2$ | 9.96 | 0.84 |
| $C_9H_{14}N_2$ | 10.71 | 0.52 |
| $C_{10}H_{14}O$ | 11.07 | 0.75 |
| $C_{11}H_{18}$ | 12.18 | 0.68 |
| $C_{12}H_6$ | 13.06 | 0.78 |
| | | |
| **151** | | |
| $C_4H_{13}N_3O_3$ | 5.79 | 0.74 |
| $C_5H_{13}NO_4$ | 6.15 | 0.96 |
| $C_6H_5N_3O_2$ | 7.79 | 0.67 |
| $C_7H_5NO_3$ | 8.14 | 0.89 |
| $C_7H_9N_3O$ | 8.89 | 0.55 |
| $C_8H_9NO_2$ | 9.25 | 0.78 |
| $C_8H_{13}N_3$ | 10.00 | 0.45 |
| $C_9H_{13}NO$ | 10.36 | 0.68 |
| $C_{10}H_{17}N$ | 11.46 | 0.60 |
| | | |
| **152** | | |
| $C_4H_{12}N_2O_4$ | 5.43 | 0.92 |
| $C_5H_4N_4O_2$ | 7.07 | 0.62 |
| $C_6H_4N_2O_3$ | 7.43 | 0.84 |
| $C_6H_8N_4O$ | 8.18 | 0.50 |
| $C_7H_4O_4$ | 7.79 | 1.06 |
| $C_7H_8N_2O_2$ | 8.53 | 0.72 |
| $C_7H_{12}N_4$ | 9.28 | 0.39 |
| $C_8H_8O_3$ | 8.89 | 0.95 |
| $C_8H_{12}N_2O$ | 9.64 | 0.62 |
| $C_9H_{12}O_2$ | 10.00 | 0.85 |
| $C_9H_{16}N_2$ | 10.74 | 0.52 |
| $C_{10}H_{16}O$ | 11.10 | 0.76 |
| $C_{11}H_4O$ | 11.99 | 0.86 |
| $C_{11}H_{20}$ | 12.21 | 0.68 |
| $C_{12}H_8$ | 13.10 | 0.79 |
| | | |
| **153** | | |
| $C_5H_3N_3O_3$ | 6.71 | 0.80 |
| $C_6H_3NO_4$ | 7.07 | 1.02 |
| $C_6H_7N_3O_2$ | 7.82 | 0.67 |
| $C_7H_7NO_3$ | 8.18 | 0.89 |
| $C_7H_{11}N_3O$ | 8.92 | 0.56 |
| $C_8H_{11}NO_2$ | 9.28 | 0.78 |
| $C_8H_{15}N_3$ | 10.03 | 0.45 |
| $C_9H_{15}NO$ | 10.39 | 0.69 |
| $C_{10}H_{19}N$ | 11.49 | 0.60 |
| $C_{11}H_7N$ | 12.38 | 0.70 |

| | M + 1 | M + 2 |
|---|---|---|
| **154** | | |
| $C_5H_6N_4O_2$ | 7.10 | 0.62 |
| $C_6H_6N_2O_3$ | 7.46 | 0.84 |
| $C_6H_{10}N_4O$ | 8.21 | 0.50 |
| $C_7H_6O_4$ | 7.82 | 1.07 |
| $C_7H_{10}N_2O_2$ | 8.57 | 0.73 |
| $C_7H_{14}N_4$ | 9.31 | 0.39 |
| $C_8H_{10}O_3$ | 8.92 | 0.95 |
| $C_8H_{14}N_2O$ | 9.67 | 0.62 |
| $C_9H_2N_2O$ | 10.56 | 0.70 |
| $C_9H_{14}O_2$ | 10.03 | 0.85 |
| $C_9H_{18}N_2$ | 10.78 | 0.53 |
| $C_{10}H_6N_2$ | 11.67 | 0.62 |
| $C_{10}H_{18}O$ | 11.13 | 0.76 |
| $C_{11}H_6O$ | 12.02 | 0.86 |
| $C_{11}H_{22}$ | 12.24 | 0.68 |
| $C_{12}H_{10}$ | 13.13 | 0.79 |
| | | |
| **155** | | |
| $C_5H_5N_3O_3$ | 6.75 | 0.80 |
| $C_6H_5NO_4$ | 7.10 | 1.02 |
| $C_6H_9N_3O_2$ | 7.85 | 0.67 |
| $C_7H_9NO_3$ | 8.21 | 0.90 |
| $C_7H_{13}N_3O$ | 8.96 | 0.56 |
| $C_8H_{13}NO_2$ | 9.31 | 0.79 |
| $C_8H_{17}N_3$ | 10.06 | 0.46 |
| $C_9H_5N_3$ | 10.95 | 0.54 |
| $C_9H_{17}NO$ | 10.42 | 0.69 |
| $C_{10}H_5NO$ | 11.31 | 0.78 |
| $C_{10}H_{21}N$ | 11.52 | 0.60 |
| $C_{11}H_9N$ | 12.41 | 0.71 |
| | | |
| **156** | | |
| $C_5H_4N_2O_4$ | 6.39 | 0.98 |
| $C_5H_8N_4O_2$ | 7.14 | 0.62 |
| $C_6H_8N_2O_3$ | 7.49 | 0.85 |
| $C_6H_{12}N_4O$ | 8.24 | 0.50 |
| $C_7H_8O_4$ | 7.85 | 1.07 |
| $C_7H_{12}N_2O_2$ | 8.60 | 0.73 |
| $C_7H_{16}N_4$ | 9.35 | 0.39 |
| $C_8H_{12}O_3$ | 8.95 | 0.96 |
| $C_8H_{16}N_2O$ | 9.70 | 0.62 |
| $C_9H_4N_2O$ | 10.59 | 0.71 |
| $C_9H_{16}O_2$ | 10.06 | 0.85 |
| $C_9H_{20}N_2$ | 10.81 | 0.53 |
| $C_{10}H_4O_2$ | 10.95 | 0.94 |
| $C_{10}H_8N_2$ | 11.70 | 0.62 |
| $C_{10}H_{20}O$ | 11.17 | 0.77 |
| $C_{11}H_8O$ | 12.05 | 0.86 |
| $C_{11}H_{24}$ | 12.27 | 0.69 |
| $C_{12}H_{12}$ | 13.16 | 0.80 |

| | M + 1 | M + 2 |
|---|---|---|
| **157** | | |
| $C_5H_7N_3O_3$ | 6.78 | 0.80 |
| $C_6H_7NO_4$ | 7.13 | 1.02 |
| $C_6H_{11}N_3O_2$ | 7.88 | 0.67 |
| $C_7H_{11}NO_3$ | 8.24 | 0.90 |
| $C_7H_{15}N_3O$ | 8.99 | 0.56 |
| $C_8H_3N_3O$ | 9.88 | 0.64 |
| $C_8H_{15}NO_2$ | 9.35 | 0.79 |
| $C_8H_{19}N_3$ | 10.09 | 0.46 |
| $C_9H_3NO_2$ | 10.23 | 0.87 |
| $C_9H_7N_3$ | 10.98 | 0.55 |
| $C_9H_{19}NO$ | 10.45 | 0.69 |
| $C_{10}H_7NO$ | 11.34 | 0.78 |
| $C_{10}H_{23}N$ | 11.56 | 0.61 |
| $C_{11}H_{11}N$ | 12.44 | 0.71 |
| | | |
| **158** | | |
| $C_5H_6N_2O_4$ | 6.42 | 0.98 |
| $C_5H_{10}N_4O_2$ | 7.17 | 0.63 |
| $C_6H_{10}N_2O_3$ | 7.52 | 0.85 |
| $C_6H_{14}N_4O$ | 8.27 | 0.50 |
| $C_7H_{10}O_4$ | 7.88 | 1.07 |
| $C_7H_{14}N_2O_2$ | 8.63 | 0.73 |
| $C_7H_{18}N_4$ | 9.38 | 0.40 |
| $C_8H_6N_4$ | 10.27 | 0.48 |
| $C_8H_{14}O_3$ | 8.99 | 0.96 |
| $C_8H_{18}N_2O$ | 9.74 | 0.63 |
| $C_9H_6N_2O$ | 10.62 | 0.71 |
| $C_9H_{18}O_2$ | 10.09 | 0.86 |
| $C_9H_{22}N_2$ | 10.84 | 0.53 |
| $C_{10}H_6O_2$ | 10.98 | 0.95 |
| $C_{10}H_{10}N_2$ | 11.73 | 0.63 |
| $C_{10}H_{22}O$ | 11.20 | 0.77 |
| $C_{11}H_{10}O$ | 12.09 | 0.87 |
| $C_{12}H_{14}$ | 13.19 | 0.80 |
| | | |
| **159** | | |
| $C_5H_9N_3O_3$ | 6.81 | 0.80 |
| $C_6H_9NO_4$ | 7.17 | 1.02 |
| $C_6H_{13}N_3O_2$ | 7.91 | 0.68 |
| $C_7H_{13}NO_3$ | 8.27 | 0.90 |
| $C_7H_{17}N_3O$ | 9.02 | 0.56 |
| $C_8H_5N_3O$ | 9.91 | 0.64 |
| $C_8H_{17}NO_2$ | 9.38 | 0.79 |
| $C_8H_{21}N_3$ | 10.13 | 0.46 |
| $C_9H_5NO_2$ | 10.27 | 0.87 |
| $C_9H_9N_3$ | 11.01 | 0.55 |
| $C_9H_{21}NO$ | 10.48 | 0.70 |
| $C_{10}H_9NO$ | 11.37 | 0.79 |
| $C_{11}H_{13}N$ | 12.48 | 0.71 |

| | M + 1 | M + 2 |
|---|---|---|
| **160** | | |
| $C_5H_8N_2O_4$ | 6.45 | 0.98 |
| $C_5H_{12}N_4O_2$ | 7.20 | 0.63 |
| $C_6H_{12}N_2O_3$ | 7.56 | 0.85 |
| $C_6H_{16}N_4O$ | 8.31 | 0.51 |
| $C_7H_4N_4O$ | 9.19 | 0.58 |
| $C_7H_{12}O_4$ | 7.91 | 1.07 |
| $C_7H_{16}N_2O_2$ | 8.66 | 0.73 |
| $C_7H_{20}N_4$ | 9.41 | 0.40 |
| $C_8H_8N_4$ | 10.30 | 0.48 |
| $C_8H_{16}O_3$ | 9.02 | 0.96 |
| $C_8H_{20}N_2O$ | 9.77 | 0.63 |
| $C_9H_9N_2O$ | 10.66 | 0.71 |
| $C_9H_{20}O_2$ | 10.12 | 0.86 |
| $C_{10}H_5O_2$ | 11.01 | 0.95 |
| $C_{10}H_{12}N_2$ | 11.76 | 0.63 |
| $C_{11}H_{12}O$ | 12.12 | 0.87 |
| $C_{12}H_{16}$ | 13.22 | 0.80 |
| | | |
| **161** | | |
| $C_5H_{11}N_3O_3$ | 6.84 | 0.80 |
| $C_6H_{11}NO_4$ | 7.20 | 1.03 |
| $C_6H_{15}N_3O_2$ | 7.95 | 0.68 |
| $C_7H_{15}NO_3$ | 8.30 | 0.90 |
| $C_7H_{19}N_3O$ | 9.05 | 0.57 |
| $C_8H_3NO_3$ | 9.19 | 0.98 |
| $C_8H_7N_3O$ | 9.94 | 0.65 |
| $C_8H_{19}NO_2$ | 9.41 | 0.80 |
| $C_9H_7NO_2$ | 10.30 | 0.88 |
| $C_9H_{11}N_3$ | 11.05 | 0.56 |
| $C_{10}H_{11}NO$ | 11.40 | 0.79 |
| $C_{11}H_{15}N$ | 12.51 | 0.72 |
| | | |
| **162** | | |
| $C_5H_{10}N_2O_4$ | 6.48 | 0.98 |
| $C_5H_{14}N_4O_2$ | 7.23 | 0.63 |
| $C_6H_{14}N_2O_3$ | 7.59 | 0.85 |
| $C_6H_{18}N_4O$ | 8.34 | 0.51 |
| $C_7H_6N_4O$ | 9.23 | 0.58 |
| $C_7H_{14}O_4$ | 7.95 | 1.08 |
| $C_7H_{18}N_2O_2$ | 8.69 | 0.74 |
| $C_8H_6N_2O_2$ | 9.58 | 0.81 |
| $C_8H_{10}N_4$ | 10.33 | 0.48 |
| $C_8H_{18}O_3$ | 9.05 | 0.96 |
| $C_9H_6O_3$ | 9.94 | 1.04 |
| $C_9H_{10}N_2O$ | 10.69 | 0.72 |
| $C_{10}H_{10}O_2$ | 11.04 | 0.95 |
| $C_{10}H_{14}N_2$ | 11.79 | 0.64 |
| $C_{11}H_{14}O$ | 12.15 | 0.87 |
| $C_{12}H_{18}$ | 13.26 | 0.81 |
| $C_{13}H_6$ | 14.14 | 0.92 |

| | $M+1$ | $M+2$ |
|---|---|---|
| **163** | | |
| $C_5H_{13}N_3O_3$ | 6.87 | 0.81 |
| $C_6H_{13}NO_4$ | 7.23 | 1.03 |
| $C_6H_{17}N_3O_2$ | 7.98 | 0.68 |
| $C_7H_5N_3O_2$ | 8.87 | 0.75 |
| $C_7H_{17}NO_3$ | 8.34 | 0.91 |
| $C_8H_5NO_3$ | 9.22 | 0.98 |
| $C_8H_9N_3O$ | 9.97 | 0.65 |
| $C_9H_9NO_2$ | 10.33 | 0.88 |
| $C_9H_{13}N_3$ | 11.08 | 0.56 |
| $C_{10}H_{13}NO$ | 11.44 | 0.80 |
| $C_{11}H_{17}N$ | 12.54 | 0.72 |
| $C_{12}H_5N$ | 13.43 | 0.83 |
| | | |
| **164** | | |
| $C_5H_{12}N_2O_4$ | 6.51 | 0.98 |
| $C_5H_{16}N_4O_2$ | 7.26 | 0.63 |
| $C_6H_4N_4O_2$ | 8.15 | 0.70 |
| $C_6H_{16}N_2O_3$ | 7.62 | 0.86 |
| $C_7H_4N_2O_3$ | 8.51 | 0.92 |
| $C_7H_8N_4O$ | 9.26 | 0.59 |
| $C_7H_{16}O_4$ | 7.98 | 1.08 |
| $C_8H_4O_4$ | 8.87 | 1.15 |
| $C_8H_8N_2O_2$ | 9.61 | 0.81 |
| $C_8H_{12}N_4$ | 10.36 | 0.49 |
| $C_9H_8O_3$ | 9.97 | 1.05 |
| $C_9H_{12}N_2O$ | 10.72 | 0.72 |
| $C_{10}H_{12}O_2$ | 11.08 | 0.96 |
| $C_{10}H_{16}N_2$ | 11.83 | 0.64 |
| $C_{11}H_{16}O$ | 12.18 | 0.88 |
| $C_{12}H_{20}$ | 13.29 | 0.81 |
| $C_{13}H_8$ | 14.18 | 0.93 |
| | | |
| **165** | | |
| $C_5H_{15}N_3O_3$ | 6.91 | 0.81 |
| $C_6H_3N_3O_3$ | 7.79 | 0.87 |
| $C_6H_{15}NO_4$ | 7.26 | 1.03 |
| $C_7H_3NO_4$ | 8.15 | 1.09 |
| $C_7H_7N_3O_2$ | 8.90 | 0.75 |
| $C_8H_7NO_3$ | 9.26 | 0.98 |
| $C_8H_{11}N_3O$ | 10.00 | 0.65 |
| $C_9H_{11}NO_2$ | 10.36 | 0.88 |
| $C_9H_{15}N_3$ | 11.11 | 0.56 |
| $C_{10}H_{15}NO$ | 11.47 | 0.80 |
| $C_{11}H_3NO$ | 12.36 | 0.90 |
| $C_{11}H_{19}N$ | 12.57 | 0.73 |
| $C_{12}H_7N$ | 13.46 | 0.84 |

| | $M+1$ | $M+2$ |
|---|---|---|
| **166** | | |
| $C_5H_{14}N_2O_4$ | 6.55 | 0.99 |
| $C_6H_6N_4O_2$ | 8.18 | 0.70 |
| $C_7H_6N_2O_3$ | 8.54 | 0.92 |
| $C_7H_{10}N_4O$ | 9.29 | 0.59 |
| $C_8H_6O_4$ | 8.90 | 1.15 |
| $C_8H_{10}N_2O_2$ | 9.65 | 0.82 |
| $C_8H_{14}N_4$ | 10.40 | 0.49 |
| $C_9H_{10}O_3$ | 10.00 | 1.05 |
| $C_9H_{14}N_2O$ | 10.75 | 0.72 |
| $C_{10}H_{14}O_2$ | 11.11 | 0.96 |
| $C_{10}H_{18}N_2$ | 11.86 | 0.64 |
| $C_{11}H_6N_2$ | 12.75 | 0.75 |
| $C_{11}H_{18}O$ | 12.21 | 0.88 |
| $C_{12}H_6O$ | 13.10 | 0.99 |
| $C_{12}H_{22}$ | 13.32 | 0.82 |
| $C_{13}H_{10}$ | 14.21 | 0.93 |
| | | |
| **167** | | |
| $C_6H_5N_3O_3$ | 7.83 | 0.87 |
| $C_7H_5NO_4$ | 8.18 | 1.10 |
| $C_7H_9N_3O_2$ | 8.93 | 0.76 |
| $C_8H_9NO_3$ | 9.29 | 0.99 |
| $C_8H_{13}N_3O$ | 10.04 | 0.66 |
| $C_9H_{13}NO_2$ | 10.39 | 0.89 |
| $C_9H_{17}N_3$ | 11.14 | 0.57 |
| $C_{10}H_5N_3$ | 12.03 | 0.66 |
| $C_{10}H_{17}NO$ | 11.50 | 0.80 |
| $C_{11}H_5NO$ | 12.39 | 0.90 |
| $C_{11}H_{21}N$ | 12.60 | 0.73 |
| $C_{12}H_9N$ | 13.49 | 0.84 |
| | | |
| **168** | | |
| $C_6H_4N_2O_4$ | 7.47 | 1.04 |
| $C_6H_8N_4O_2$ | 8.22 | 0.70 |
| $C_7H_8N_2O_3$ | 8.57 | 0.93 |
| $C_7H_{12}N_4O$ | 9.32 | 0.59 |
| $C_8H_8O_4$ | 8.93 | 1.15 |
| $C_8H_{12}N_2O_2$ | 9.68 | 0.82 |
| $C_8H_{16}N_4$ | 10.43 | 0.49 |
| $C_9H_4N_4$ | 11.32 | 0.58 |
| $C_9H_{12}O_3$ | 10.04 | 1.05 |
| $C_9H_{16}N_2O$ | 10.78 | 0.73 |
| $C_{10}H_{16}O_2$ | 11.14 | 0.96 |
| $C_{10}H_{20}N_2$ | 11.89 | 0.65 |
| $C_{11}H_8N_2$ | 12.78 | 0.75 |
| $C_{11}H_{20}O$ | 12.25 | 0.89 |
| $C_{12}H_8O$ | 13.13 | 0.99 |
| $C_{12}H_{24}$ | 13.35 | 0.82 |
| $C_{13}H_{12}$ | 14.24 | 0.94 |

| | $M+1$ | $M+2$ |
|---|---|---|
| **169** | | |
| $C_6H_7N_3O_3$ | 7.86 | 0.87 |
| $C_7H_7NO_4$ | 8.21 | 1.10 |
| $C_7H_{11}N_3O_2$ | 8.96 | 0.76 |
| $C_8H_{11}NO_3$ | 9.32 | 0.99 |
| $C_8H_{15}N_3O$ | 10.07 | 0.66 |
| $C_9H_3N_3O$ | 10.96 | 0.75 |
| $C_9H_{15}NO_2$ | 10.43 | 0.89 |
| $C_9H_{19}N_3$ | 11.17 | 0.57 |
| $C_{10}H_7N_3$ | 12.06 | 0.67 |
| $C_{10}H_{19}NO$ | 11.53 | 0.81 |
| $C_{11}H_7NO$ | 12.42 | 0.91 |
| $C_{11}H_{23}N$ | 12.64 | 0.73 |
| $C_{12}H_{11}N$ | 13.53 | 0.84 |
| | | |
| **170** | | |
| $C_6H_6N_2O_4$ | 7.50 | 1.05 |
| $C_6H_{10}N_4O_2$ | 8.25 | 0.70 |
| $C_7H_{10}N_2O_3$ | 8.60 | 0.93 |
| $C_7H_{14}N_4O$ | 9.35 | 0.59 |
| $C_8H_{10}O_4$ | 8.96 | 1.16 |
| $C_8H_{14}N_2O_2$ | 9.71 | 0.82 |
| $C_8H_{18}N_4$ | 10.46 | 0.50 |
| $C_9H_2N_2O_2$ | 10.60 | 0.91 |
| $C_9H_6N_4$ | 11.35 | 0.59 |
| $C_9H_{14}O_3$ | 10.07 | 1.06 |
| $C_9H_{18}N_2O$ | 10.82 | 0.73 |
| $C_{10}H_6N_2O$ | 11.70 | 0.83 |
| $C_{10}H_{18}O_2$ | 11.17 | 0.97 |
| $C_{10}H_{22}N_2$ | 11.92 | 0.65 |
| $C_{11}H_6O_2$ | 12.06 | 1.06 |
| $C_{11}H_{10}N_2$ | 12.81 | 0.75 |
| $C_{11}H_{22}O$ | 12.28 | 0.89 |
| $C_{12}H_{10}O$ | 13.17 | 1.00 |
| $C_{12}H_{26}$ | 13.38 | 0.83 |
| $C_{13}H_{14}$ | 14.27 | 0.94 |
| | | |
| **171** | | |
| $C_6H_9N_3O_3$ | 7.89 | 0.88 |
| $C_7H_9NO_4$ | 8.25 | 1.10 |
| $C_7H_{13}N_3O_2$ | 9.00 | 0.76 |
| $C_8H_{13}NO_3$ | 9.35 | 0.99 |
| $C_8H_{17}N_3O$ | 10.10 | 0.66 |
| $C_9H_5N_3O$ | 10.99 | 0.75 |
| $C_9H_{17}NO_2$ | 10.46 | 0.89 |
| $C_9H_{21}N_3$ | 11.21 | 0.57 |
| $C_{10}H_9N_3$ | 12.09 | 0.67 |
| $C_{10}H_{21}NO$ | 11.56 | 0.81 |
| $C_{11}H_9NO$ | 12.45 | 0.91 |
| $C_{11}H_{25}N$ | 12.67 | 0.74 |
| $C_{12}H_{13}N$ | 13.56 | 0.85 |

|  | $M+1$ | $M+2$ |
|---|---|---|
| **172** | | |
| $C_6H_8N_2O_4$ | 7.53 | 1.05 |
| $C_6H_{12}N_4O_2$ | 8.28 | 0.71 |
| $C_7H_{12}N_2O_3$ | 8.64 | 0.93 |
| $C_7H_{16}N_4O$ | 9.39 | 0.60 |
| $C_8H_{12}O_4$ | 8.99 | 1.16 |
| $C_8H_{16}N_2O_2$ | 9.74 | 0.83 |
| $C_8H_{20}N_4$ | 10.49 | 0.50 |
| $C_9H_4N_2O_2$ | 10.63 | 0.91 |
| $C_9H_8N_4$ | 11.38 | 0.59 |
| $C_9H_{16}O_3$ | 10.10 | 1.06 |
| $C_9H_{20}N_2O$ | 10.85 | 0.73 |
| $C_{10}H_8N_2O$ | 11.74 | 0.83 |
| $C_{10}H_{20}O_2$ | 11.20 | 0.97 |
| $C_{10}H_{24}N_2$ | 11.95 | 0.65 |
| $C_{11}H_8O_2$ | 12.09 | 1.07 |
| $C_{11}H_{12}N_2$ | 12.84 | 0.76 |
| $C_{11}H_{24}O$ | 12.31 | 0.89 |
| $C_{12}H_{12}O$ | 13.20 | 1.00 |
| $C_{13}H_{16}$ | 14.30 | 0.95 |

|  | $M+1$ | $M+2$ |
|---|---|---|
| **173** | | |
| $C_6H_{11}N_3O_3$ | 7.92 | 0.88 |
| $C_7H_{11}NO_4$ | 8.28 | 1.10 |
| $C_7H_{15}N_3O_2$ | 9.03 | 0.77 |
| $C_8H_{15}NO_3$ | 9.38 | 0.99 |
| $C_8H_{19}N_3O$ | 10.13 | 0.66 |
| $C_9H_3NO_3$ | 10.27 | 1.08 |
| $C_9H_7N_3O$ | 11.02 | 0.75 |
| $C_9H_{19}NO_2$ | 10.49 | 0.90 |
| $C_9H_{23}N_3$ | 11.24 | 0.58 |
| $C_{10}H_7NO_2$ | 11.38 | 0.99 |
| $C_{10}H_{11}N_3$ | 12.13 | 0.67 |
| $C_{10}H_{23}NO$ | 11.60 | 0.81 |
| $C_{11}H_{11}NO$ | 12.48 | 0.91 |
| $C_{12}H_{15}N$ | 13.59 | 0.85 |

|  | $M+1$ | $M+2$ |
|---|---|---|
| **174** | | |
| $C_6H_{10}N_2O_4$ | 7.56 | 1.05 |
| $C_6H_{14}N_4O_2$ | 8.31 | 0.71 |
| $C_7H_{14}N_2O_3$ | 8.67 | 0.93 |
| $C_7H_{19}N_4O$ | 9.42 | 0.60 |
| $C_8H_6N_4O$ | 10.31 | 0.68 |
| $C_8H_{14}O_4$ | 9.03 | 1.16 |
| $C_8H_{18}N_2O_2$ | 9.77 | 0.83 |
| $C_8H_{22}N_4$ | 10.52 | 0.50 |
| $C_9H_6N_2O_2$ | 10.66 | 0.92 |
| $C_9H_{10}N_4$ | 11.41 | 0.60 |
| $C_9H_{18}O_3$ | 10.13 | 1.06 |
| $C_9H_{22}N_2O$ | 10.88 | 0.74 |
| $C_{10}H_6O_3$ | 11.02 | 1.15 |

|  | $M+1$ | $M+2$ |
|---|---|---|
| $C_{10}H_{10}N_2O$ | 11.77 | 0.83 |
| $C_{10}H_{22}O_2$ | 11.24 | 0.97 |
| $C_{11}H_{10}O_2$ | 12.13 | 1.07 |
| $C_{11}H_{14}N_2$ | 12.87 | 0.76 |
| $C_{12}H_{14}O$ | 13.23 | 1.01 |
| $C_{13}H_{18}$ | 14.34 | 0.95 |
| $C_{14}H_6$ | 15.22 | 1.08 |

|  | $M+1$ | $M+2$ |
|---|---|---|
| **175** | | |
| $C_6H_{13}N_3O_3$ | 7.95 | 0.88 |
| $C_7H_{13}NO_4$ | 8.31 | 1.11 |
| $C_7H_{17}N_3O_2$ | 9.06 | 0.77 |
| $C_8H_5N_3O_2$ | 9.95 | 0.85 |
| $C_8H_{17}NO_3$ | 9.42 | 1.00 |
| $C_8H_{21}N_3O$ | 10.16 | 0.67 |
| $C_9H_5NO_3$ | 10.30 | 1.08 |
| $C_9H_9N_3O$ | 11.05 | 0.76 |
| $C_9H_{21}NO_2$ | 10.52 | 0.90 |
| $C_{10}H_9NO_2$ | 11.41 | 0.99 |
| $C_{10}H_{13}N_3$ | 12.16 | 0.68 |
| $C_{11}H_{13}NO$ | 12.52 | 0.92 |
| $C_{12}H_{17}N$ | 13.62 | 0.86 |

|  | $M+1$ | $M+2$ |
|---|---|---|
| **176** | | |
| $C_6H_{12}N_2O_4$ | 7.60 | 1.05 |
| $C_6H_{16}N_4O_2$ | 8.34 | 0.71 |
| $C_7H_{16}N_2O_3$ | 8.70 | 0.94 |
| $C_7H_{20}N_4O$ | 9.45 | 0.60 |
| $C_8H_4N_2O_3$ | 9.59 | 1.01 |
| $C_8H_8N_4O$ | 10.34 | 0.69 |
| $C_8H_{16}O_4$ | 9.06 | 1.17 |
| $C_8H_{20}N_2O_2$ | 9.81 | 0.83 |
| $C_9H_4O_4$ | 9.95 | 1.24 |
| $C_9H_8N_2O_2$ | 10.70 | 0.92 |
| $C_9H_{12}N_4$ | 11.44 | 0.60 |
| $C_9H_{20}O_3$ | 10.16 | 1.07 |
| $C_{10}H_8O_3$ | 11.05 | 1.15 |
| $C_{10}H_{12}N_2O$ | 11.80 | 0.84 |
| $C_{11}H_{12}O_2$ | 12.16 | 1.08 |
| $C_{11}H_{16}N_2$ | 12.91 | 0.77 |
| $C_{12}H_4N_2$ | 13.79 | 0.88 |
| $C_{12}H_{16}O$ | 13.26 | 1.01 |
| $C_{13}H_{20}$ | 14.37 | 0.96 |
| $C_{14}H_8$ | 15.26 | 1.08 |

|  | $M+1$ | $M+2$ |
|---|---|---|
| **177** | | |
| $C_6H_{15}N_3O_3$ | 7.99 | 0.88 |
| $C_7H_{15}NO_4$ | 8.34 | 1.11 |
| $C_7H_{19}N_3O_2$ | 9.09 | 0.77 |
| $C_8H_3NO_4$ | 9.23 | 1.18 |
| $C_8H_7N_3O_2$ | 9.98 | 0.85 |

|  | $M+1$ | $M+2$ |
|---|---|---|
| $C_8H_{19}NO_3$ | 9.45 | 1.00 |
| $C_9H_7NO_3$ | 10.34 | 1.08 |
| $C_9H_{11}N_3O$ | 11.09 | 0.76 |
| $C_{10}H_{11}NO_2$ | 11.44 | 1.00 |
| $C_{10}H_{15}N_3$ | 12.19 | 0.68 |
| $C_{11}H_3N_3$ | 13.08 | 0.79 |
| $C_{11}H_{15}NO$ | 12.55 | 0.92 |
| $C_{12}H_3NO$ | 13.44 | 1.03 |
| $C_{12}H_{19}N$ | 13.65 | 0.86 |
| $C_{13}H_7N$ | 14.54 | 0.98 |

|  | $M+1$ | $M+2$ |
|---|---|---|
| **178** | | |
| $C_6H_{14}N_2O_4$ | 7.63 | 1.06 |
| $C_6H_{18}N_4O_2$ | 8.38 | 0.71 |
| $C_7H_6N_4O_2$ | 9.26 | 0.79 |
| $C_7H_{18}N_2O_3$ | 8.73 | 0.94 |
| $C_8H_6N_2O_3$ | 9.62 | 1.02 |
| $C_8H_{10}N_4O$ | 10.37 | 0.69 |
| $C_8H_{18}O_4$ | 9.09 | 1.17 |
| $C_9H_6O_4$ | 9.98 | 1.25 |
| $C_9H_{10}N_2O_2$ | 10.73 | 0.92 |
| $C_9H_{14}N_4$ | 11.48 | 0.60 |
| $C_{10}H_{10}O_3$ | 11.08 | 1.16 |
| $C_{10}H_{14}N_2O$ | 11.83 | 0.84 |
| $C_{11}H_{14}O_2$ | 12.19 | 1.08 |
| $C_{11}H_{18}N_2$ | 12.94 | 0.77 |
| $C_{12}H_6N_2$ | 13.83 | 0.88 |
| $C_{12}H_{18}O$ | 13.29 | 1.01 |
| $C_{13}H_6O$ | 14.18 | 1.13 |
| $C_{13}H_{22}$ | 14.40 | 0.96 |
| $C_{14}H_{10}$ | 15.29 | 1.09 |

|  | $M+1$ | $M+2$ |
|---|---|---|
| **179** | | |
| $C_6H_{17}N_3O_3$ | 8.02 | 0.89 |
| $C_7H_5N_3O_3$ | 8.91 | 0.95 |
| $C_7H_{17}NO_4$ | 8.37 | 1.11 |
| $C_8H_5NO_4$ | 9.26 | 1.18 |
| $C_8H_9N_3O_2$ | 10.01 | 0.85 |
| $C_9H_9NO_3$ | 10.37 | 1.09 |
| $C_9H_{13}N_3O$ | 11.12 | 0.76 |
| $C_{10}H_{11}O_3$ | 11.10 | 1.16 |
| $C_{10}H_{13}NO_2$ | 11.47 | 1.00 |
| $C_{10}H_{17}N_3$ | 12.22 | 0.69 |
| $C_{11}H_{17}NO$ | 12.58 | 0.93 |
| $C_{12}H_5NO$ | 13.47 | 1.04 |
| $C_{12}H_{21}N$ | 13.69 | 0.87 |
| $C_{13}H_9N$ | 14.57 | 0.99 |

|  | $M+1$ | $M+2$ |
|---|---|---|
| **180** | | |
| $C_6H_{16}N_2O_4$ | 7.66 | 1.06 |
| $C_7H_4N_2O_4$ | 8.55 | 1.12 |

| | M + 1 | M + 2 |
|---|---|---|
| $C_7H_8N_4O_2$ | 9.30 | 0.79 |
| $C_8H_8N_2O_3$ | 9.65 | 1.02 |
| $C_8H_{12}N_4O$ | 10.40 | 0.69 |
| $C_9H_8O_4$ | 10.01 | 1.25 |
| $C_9H_{12}N_2O_2$ | 10.76 | 0.93 |
| $C_9H_{16}N_4$ | 11.51 | 0.61 |
| $C_{10}H_{12}O_3$ | 11.12 | 1.16 |
| $C_{10}H_{16}N_2O$ | 11.86 | 0.84 |
| $C_{11}H_{16}O_2$ | 12.22 | 1.08 |
| $C_{11}H_{20}N_2$ | 12.97 | 0.78 |
| $C_{12}H_4O_2$ | 13.11 | 1.19 |
| $C_{12}H_8N_2$ | 13.86 | 0.89 |
| $C_{12}H_{20}O$ | 13.33 | 1.02 |
| $C_{13}H_8O$ | 14.22 | 1.13 |
| $C_{13}H_{24}$ | 14.43 | 0.97 |
| $C_{14}H_{12}$ | 15.32 | 1.09 |

**181**

| | M + 1 | M + 2 |
|---|---|---|
| $C_7H_7N_3O_3$ | 8.94 | 0.96 |
| $C_8H_7NO_4$ | 9.30 | 1.19 |
| $C_8H_{11}N_3O_2$ | 10.04 | 0.86 |
| $C_9H_{11}NO_3$ | 10.40 | 1.09 |
| $C_9H_{15}N_3O$ | 11.15 | 0.77 |
| $C_{10}H_{15}NO_2$ | 11.51 | 1.00 |
| $C_{10}H_{19}N_3$ | 12.25 | 0.69 |
| $C_{11}H_7N_3$ | 13.14 | 0.80 |
| $C_{11}H_{19}NO$ | 12.61 | 0.93 |
| $C_{12}H_7NO$ | 13.50 | 1.04 |
| $C_{12}H_{23}N$ | 13.72 | 0.87 |
| $C_{13}H_{11}N$ | 14.61 | 0.99 |

**182**

| | M + 1 | M + 2 |
|---|---|---|
| $C_7H_6N_2O_4$ | 8.58 | 1.13 |
| $C_7H_{10}N_4O_2$ | 9.33 | 0.79 |
| $C_8H_{10}N_2O_3$ | 9.69 | 1.02 |
| $C_8H_{14}N_4O$ | 10.43 | 0.70 |
| $C_9H_{10}O_4$ | 10.04 | 1.25 |
| $C_9H_{14}N_2O_2$ | 10.79 | 0.93 |
| $C_9H_{18}N_4$ | 11.54 | 0.61 |
| $C_{10}H_6N_4$ | 12.43 | 0.71 |
| $C_{10}H_{14}O_3$ | 11.15 | 1.16 |
| $C_{10}H_{18}N_2O$ | 11.90 | 0.85 |
| $C_{11}H_6N_2O$ | 12.79 | 0.95 |
| $C_{11}H_{18}O_2$ | 12.25 | 1.09 |
| $C_{11}H_{22}N_2$ | 13.00 | 0.78 |
| $C_{12}H_6O_2$ | 13.14 | 1.19 |
| $C_{12}H_{10}N_2$ | 13.89 | 0.89 |
| $C_{12}H_{22}O$ | 13.36 | 1.02 |
| $C_{13}H_{10}O$ | 14.25 | 1.14 |
| $C_{13}H_{26}$ | 14.46 | 0.97 |
| $C_{14}H_{14}$ | 15.35 | 1.10 |

**183**

| | M + 1 | M + 2 |
|---|---|---|
| $C_7H_9N_3O_3$ | 8.97 | 0.96 |
| $C_8H_9NO_4$ | 9.33 | 1.19 |
| $C_8H_{13}N_3O_2$ | 10.08 | 0.86 |
| $C_9H_{13}NO_3$ | 10.43 | 1.09 |
| $C_9H_{17}N_3O$ | 11.18 | 0.77 |
| $C_{10}H_5N_3O$ | 12.07 | 0.87 |
| $C_{10}H_{17}NO_2$ | 11.54 | 1.01 |
| $C_{10}H_{21}N_3$ | 12.29 | 0.69 |
| $C_{11}H_5NO_2$ | 12.43 | 1.11 |
| $C_{11}H_9N_3$ | 13.18 | 0.80 |
| $C_{11}H_{21}NO$ | 12.64 | 0.93 |
| $C_{12}H_9NO$ | 13.53 | 1.05 |
| $C_{12}H_{25}N$ | 13.75 | 0.87 |
| $C_{13}H_{13}N$ | 14.64 | 0.99 |

**184**

| | M + 1 | M + 2 |
|---|---|---|
| $C_7H_8N_2O_4$ | 8.61 | 1.13 |
| $C_7H_{12}N_4O_2$ | 9.36 | 0.80 |
| $C_8H_{12}N_2O_3$ | 9.72 | 1.03 |
| $C_8H_{16}N_4O$ | 10.47 | 0.70 |
| $C_9H_{12}O_4$ | 10.07 | 1.26 |
| $C_9H_{16}N_2O_2$ | 10.82 | 0.93 |
| $C_9H_{20}N_4$ | 11.57 | 0.61 |
| $C_{10}H_8N_4$ | 12.46 | 0.71 |
| $C_{10}H_{16}O_3$ | 11.18 | 1.17 |
| $C_{10}H_{20}N_2O$ | 11.93 | 0.85 |
| $C_{11}H_8N_2O$ | 12.82 | 0.96 |
| $C_{11}H_{20}O_2$ | 12.29 | 1.09 |
| $C_{11}H_{24}N_2$ | 13.03 | 0.78 |
| $C_{12}H_8O_2$ | 13.17 | 1.20 |
| $C_{12}H_{12}N_2$ | 13.92 | 0.90 |
| $C_{12}H_{24}O$ | 13.39 | 1.03 |
| $C_{13}H_{12}O$ | 14.28 | 1.14 |
| $C_{13}H_{28}$ | 14.50 | 0.97 |
| $C_{14}H_{16}$ | 15.38 | 1.10 |

**185**

| | M + 1 | M + 2 |
|---|---|---|
| $C_7H_{11}N_3O_3$ | 9.00 | 0.96 |
| $C_8H_{11}NO_4$ | 9.36 | 1.19 |
| $C_8H_{15}N_3O_2$ | 10.11 | 0.86 |
| $C_9H_{15}NO_3$ | 10.46 | 1.10 |
| $C_9H_{19}N_3O$ | 11.21 | 0.77 |
| $C_{10}H_7N_3O$ | 12.10 | 0.87 |
| $C_{10}H_{19}NO_2$ | 11.57 | 1.01 |
| $C_{10}H_{23}N_3$ | 12.32 | 0.70 |
| $C_{11}H_7NO_2$ | 12.46 | 1.11 |
| $C_{11}H_{11}N_3$ | 13.21 | 0.81 |
| $C_{11}H_{23}NO$ | 12.68 | 0.94 |
| $C_{12}H_{11}NO$ | 13.56 | 1.05 |
| $C_{12}H_{27}N$ | 13.78 | 0.88 |
| $C_{13}H_{15}N$ | 14.67 | 1.00 |

**186**

| | M + 1 | M + 2 |
|---|---|---|
| $C_7H_{10}N_2O_4$ | 8.64 | 1.13 |
| $C_7H_{14}N_4O_2$ | 9.39 | 0.80 |
| $C_8H_{14}N_2O_3$ | 9.75 | 1.03 |
| $C_8H_{18}N_4O$ | 10.50 | 0.70 |
| $C_9H_6N_4O$ | 11.39 | 0.79 |
| $C_9H_{14}O_4$ | 10.11 | 1.26 |
| $C_9H_{18}N_2O_2$ | 10.86 | 0.94 |
| $C_9H_{22}N_4$ | 11.60 | 0.62 |
| $C_{10}H_6N_2O_2$ | 11.74 | 1.03 |
| $C_{10}H_{10}N_4$ | 12.49 | 0.72 |
| $C_{10}H_{18}O_3$ | 11.21 | 1.17 |
| $C_{10}H_{22}N_2O$ | 11.96 | 0.86 |
| $C_{11}H_6O_3$ | 12.10 | 1.27 |
| $C_{11}H_{10}N_2O$ | 12.85 | 0.96 |
| $C_{11}H_{22}O_2$ | 12.32 | 1.10 |
| $C_{11}H_{26}N_2$ | 13.07 | 0.79 |
| $C_{12}H_{10}O_2$ | 13.21 | 1.20 |
| $C_{12}H_{14}N_2$ | 13.95 | 0.90 |
| $C_{12}H_{26}O$ | 13.42 | 1.03 |
| $C_{13}H_2N_2$ | 14.84 | 1.02 |
| $C_{13}H_{14}O$ | 14.31 | 1.15 |
| $C_{14}H_2O$ | 15.20 | 1.27 |
| $C_{14}H_{18}$ | 15.42 | 1.11 |
| $C_{15}H_6$ | 16.31 | 1.24 |

**187**

| | M + 1 | M + 2 |
|---|---|---|
| $C_7H_{13}N_3O_3$ | 9.03 | 0.97 |
| $C_8H_{13}NO_4$ | 9.39 | 1.20 |
| $C_8H_{17}N_3O_2$ | 10.14 | 0.87 |
| $C_9H_5N_3O_2$ | 11.03 | 0.95 |
| $C_9H_{17}NO_3$ | 10.50 | 1.10 |
| $C_9H_{21}N_3O$ | 11.25 | 0.78 |
| $C_{10}H_9N_3O$ | 12.13 | 0.88 |
| $C_{10}H_{21}NO_2$ | 11.60 | 1.01 |
| $C_{10}H_{25}N_3$ | 12.35 | 0.70 |
| $C_{11}H_9NO_2$ | 12.49 | 1.12 |
| $C_{11}H_{13}N_3$ | 13.24 | 0.81 |
| $C_{11}H_{25}NO$ | 12.71 | 0.94 |
| $C_{12}H_{13}NO$ | 13.60 | 1.05 |
| $C_{13}H_{17}N$ | 14.70 | 1.00 |
| $C_{14}H_5N$ | 15.59 | 1.13 |

**188**

| | M + 1 | M + 2 |
|---|---|---|
| $C_7H_{12}N_2O_4$ | 8.68 | 1.14 |
| $C_7H_{16}N_4O_2$ | 9.42 | 0.80 |
| $C_8H_4N_4O_2$ | 10.31 | 0.88 |
| $C_8H_{16}N_2O_3$ | 9.78 | 1.03 |
| $C_8H_{20}N_4O$ | 10.53 | 0.71 |
| $C_9H_4N_2O_3$ | 10.67 | 1.12 |
| $C_9H_8N_4O$ | 11.42 | 0.80 |

| | $M+1$ | $M+2$ |
|---|---|---|
| $C_9H_{16}O_4$ | 10.14 | 1.26 |
| $C_9H_{20}N_2O_2$ | 10.89 | 0.94 |
| $C_9H_{24}N_4$ | 11.64 | 0.62 |
| $C_{10}H_4O_4$ | 11.03 | 1.35 |
| $C_{10}H_8N_2O_2$ | 11.78 | 1.03 |
| $C_{10}H_{12}N_4$ | 12.52 | 0.72 |
| $C_{10}H_{20}O_3$ | 11.24 | 1.18 |
| $C_{10}H_{24}N_2O$ | 11.99 | 0.86 |
| $C_{11}H_8O_3$ | 12.13 | 1.27 |
| $C_{11}H_{12}N_2O$ | 12.88 | 0.96 |
| $C_{11}H_{24}O_2$ | 12.35 | 1.10 |
| $C_{12}H_{12}O_2$ | 13.24 | 1.21 |
| $C_{12}H_{16}N_2$ | 13.99 | 0.91 |
| $C_{13}H_{16}O$ | 14.34 | 1.15 |
| $C_{14}H_{20}$ | 15.45 | 1.11 |
| $C_{15}H_8$ | 16.34 | 1.25 |

**189**

| | $M+1$ | $M+2$ |
|---|---|---|
| $C_7H_{15}N_3O_3$ | 9.07 | 0.97 |
| $C_8H_{15}NO_4$ | 9.42 | 1.20 |
| $C_8H_{19}N_3O_2$ | 10.17 | 0.87 |
| $C_9H_7N_3O_2$ | 11.06 | 0.96 |
| $C_9H_{19}NO_3$ | 10.53 | 1.10 |
| $C_9H_{23}N_3O$ | 11.28 | 0.78 |
| $C_{10}H_7NO_3$ | 11.42 | 1.19 |
| $C_{10}H_{11}N_3O$ | 12.17 | 0.88 |
| $C_{10}H_{23}NO_2$ | 11.63 | 1.02 |
| $C_{11}H_{11}NO_2$ | 12.52 | 1.12 |
| $C_{11}H_{15}N_3$ | 13.27 | 0.81 |
| $C_{12}H_{15}NO$ | 13.63 | 1.06 |
| $C_{13}H_{19}N$ | 14.73 | 1.01 |
| $C_{14}H_7N$ | 15.62 | 1.14 |

**190**

| | $M+1$ | $M+2$ |
|---|---|---|
| $C_7H_{14}N_2O_4$ | 8.71 | 1.14 |
| $C_7H_{18}N_4O_2$ | 9.46 | 0.80 |
| $C_8H_2N_2O_4$ | 9.60 | 1.21 |
| $C_8H_6N_4O_2$ | 10.35 | 0.89 |
| $C_8H_{18}N_2O_3$ | 9.81 | 1.03 |
| $C_8H_{22}N_4O$ | 10.56 | 0.71 |
| $C_9H_6N_2O_3$ | 10.70 | 1.12 |
| $C_9H_{10}N_4O$ | 11.45 | 0.80 |
| $C_9H_{18}O_4$ | 10.17 | 1.27 |
| $C_9H_{22}N_2O_2$ | 10.92 | 0.94 |
| $C_{10}H_6O_4$ | 11.06 | 1.35 |
| $C_{10}H_{10}N_2O_2$ | 11.81 | 1.03 |
| $C_{10}H_{14}N_4$ | 12.56 | 0.73 |
| $C_{10}H_{22}O_3$ | 11.28 | 1.18 |
| $C_{11}H_{10}O_3$ | 12.16 | 1.28 |
| $C_{11}H_{14}N_2O$ | 12.91 | 0.97 |

| | $M+1$ | $M+2$ |
|---|---|---|
| $C_{12}H_{14}O_2$ | 13.27 | 1.21 |
| $C_{12}H_{18}N_2$ | 14.02 | 0.91 |
| $C_{13}H_6N_2$ | 14.91 | 1.03 |
| $C_{13}H_{18}O$ | 14.38 | 1.16 |
| $C_{14}H_6O$ | 15.26 | 1.28 |
| $C_{14}H_{22}$ | 15.48 | 1.12 |
| $C_{15}H_{10}$ | 16.37 | 1.25 |

**191**

| | $M+1$ | $M+2$ |
|---|---|---|
| $C_7H_{17}N_3O_3$ | 9.10 | 0.97 |
| $C_8H_5N_3O_3$ | 9.99 | 1.05 |
| $C_8H_{17}NO_4$ | 9.46 | 1.20 |
| $C_8H_{21}N_3O_2$ | 10.20 | 0.87 |
| $C_9H_5NO_4$ | 10.34 | 1.28 |
| $C_9H_9N_3O_2$ | 11.09 | 0.96 |
| $C_9H_{21}NO_3$ | 10.56 | 1.11 |
| $C_{10}H_9NO_3$ | 11.45 | 1.20 |
| $C_{10}H_{13}N_3O$ | 12.20 | 0.88 |
| $C_{11}H_{13}NO_2$ | 12.55 | 1.12 |
| $C_{11}H_{17}N_3$ | 13.30 | 0.82 |
| $C_{12}H_{17}NO$ | 13.66 | 1.06 |
| $C_{13}H_{21}N$ | 14.77 | 1.01 |
| $C_{14}H_9N$ | 15.65 | 1.14 |

**192**

| | $M+1$ | $M+2$ |
|---|---|---|
| $C_7H_{16}N_2O_4$ | 8.74 | 1.14 |
| $C_7H_{20}N_4O_2$ | 9.49 | 0.81 |
| $C_8H_4N_2O_4$ | 9.63 | 1.22 |
| $C_8H_8N_4O_2$ | 10.38 | 0.89 |
| $C_8H_{20}N_2O_3$ | 9.85 | 1.04 |
| $C_9H_8N_2O_3$ | 10.73 | 1.12 |
| $C_9H_{12}N_4O$ | 11.48 | 0.80 |
| $C_9H_{20}O_4$ | 10.20 | 1.27 |
| $C_{10}H_8O_4$ | 11.09 | 1.36 |
| $C_{10}H_{12}N_2O_2$ | 11.84 | 1.04 |
| $C_{10}H_{16}N_4$ | 12.59 | 0.73 |
| $C_{11}H_{12}O_3$ | 12.20 | 1.28 |
| $C_{11}H_{16}N_2O$ | 12.95 | 0.97 |
| $C_{12}H_{16}O_2$ | 13.30 | 1.22 |
| $C_{12}H_{20}N_2$ | 14.05 | 0.92 |
| $C_{13}H_8N_2$ | 14.94 | 1.04 |
| $C_{13}H_{20}O$ | 14.41 | 1.16 |
| $C_{14}H_8O$ | 15.30 | 1.29 |
| $C_{14}H_{24}$ | 15.51 | 1.12 |
| $C_{15}H_{12}$ | 16.40 | 1.26 |

**193**

| | $M+1$ | $M+2$ |
|---|---|---|
| $C_7H_{19}N_3O_3$ | 9.13 | 0.98 |
| $C_8H_7N_3O_3$ | 10.02 | 1.05 |

| | $M+1$ | $M+2$ |
|---|---|---|
| $C_8H_{19}NO_4$ | 9.49 | 1.20 |
| $C_9H_7NO_4$ | 10.38 | 1.29 |
| $C_9H_{11}N_3O_2$ | 11.12 | 0.96 |
| $C_{10}H_{11}NO_3$ | 11.48 | 1.20 |
| $C_{10}H_{15}N_3O$ | 12.23 | 0.89 |
| $C_{11}H_3N_3O$ | 13.12 | 0.99 |
| $C_{11}H_{15}NO_2$ | 12.59 | 1.13 |
| $C_{11}H_{19}N_3$ | 13.34 | 0.82 |
| $C_{12}H_3NO_2$ | 13.48 | 1.24 |
| $C_{12}H_7N_3$ | 14.22 | 0.94 |
| $C_{12}H_{19}NO$ | 13.69 | 1.07 |
| $C_{13}H_7NO$ | 14.58 | 1.19 |
| $C_{13}H_{23}N$ | 14.80 | 1.02 |
| $C_{14}H_{11}N$ | 15.69 | 1.15 |

**194**

| | $M+1$ | $M+2$ |
|---|---|---|
| $C_7H_{18}N_2O_4$ | 8.77 | 1.14 |
| $C_8H_6N_2O_4$ | 9.66 | 1.22 |
| $C_8H_{10}N_4O_2$ | 10.41 | 0.89 |
| $C_9H_{10}N_2O_3$ | 10.77 | 1.13 |
| $C_9H_{14}N_4O$ | 11.51 | 0.81 |
| $C_{10}H_2N_4O$ | 12.40 | 0.91 |
| $C_{10}H_{10}O_4$ | 11.12 | 1.36 |
| $C_{10}H_{14}N_2O_2$ | 11.87 | 1.05 |
| $C_{10}H_{18}N_4$ | 12.62 | 0.74 |
| $C_{11}H_6N_4$ | 13.51 | 0.85 |
| $C_{11}H_{14}O_3$ | 12.23 | 1.28 |
| $C_{11}H_{18}N_2O$ | 12.98 | 0.98 |
| $C_{12}H_6N_2O$ | 13.87 | 1.09 |
| $C_{12}H_{18}O_2$ | 13.33 | 1.22 |
| $C_{12}H_{22}N_2$ | 14.08 | 0.92 |
| $C_{13}H_6O_2$ | 14.22 | 1.34 |
| $C_{13}H_{10}N_2$ | 14.97 | 1.04 |
| $C_{13}H_{22}O$ | 14.44 | 1.17 |
| $C_{14}H_{10}O$ | 15.33 | 1.29 |
| $C_{14}H_{26}$ | 15.54 | 1.13 |
| $C_{15}H_{14}$ | 16.43 | 1.26 |

**195**

| | $M+1$ | $M+2$ |
|---|---|---|
| $C_8H_9N_3O_3$ | 10.05 | 1.06 |
| $C_9H_9NO_4$ | 10.41 | 1.29 |
| $C_9H_{13}N_3O_2$ | 11.16 | 0.97 |
| $C_{10}H_{13}NO_3$ | 11.51 | 1.21 |
| $C_{10}H_{17}N_3O$ | 12.26 | 0.89 |
| $C_{11}H_5N_3O$ | 13.15 | 1.00 |
| $C_{11}H_{17}NO_2$ | 12.62 | 1.13 |
| $C_{11}H_{21}N_3$ | 13.37 | 0.83 |
| $C_{12}H_5NO_2$ | 13.51 | 1.24 |
| $C_{12}H_9N_3$ | 14.26 | 0.94 |
| $C_{12}H_{21}NO$ | 13.72 | 1.07 |

| | M + 1 | M + 2 |
|---|---|---|
| $C_{13}H_9NO$ | 14.61 | 1.19 |
| $C_{13}H_{25}N$ | 14.83 | 1.02 |
| $C_{14}H_{13}N$ | 15.72 | 1.15 |
| | | |
| **196** | | |
| $C_8H_8N_2O_4$ | 9.69 | 1.22 |
| $C_8H_{12}N_4O_2$ | 10.44 | 0.90 |
| $C_9H_{12}N_2O_3$ | 10.80 | 1.13 |
| $C_9H_{16}N_4O$ | 11.55 | 0.81 |
| $C_{10}H_{12}O_4$ | 11.15 | 1.37 |
| $C_{10}H_{16}N_2O_2$ | 11.90 | 1.05 |
| $C_{10}H_{20}N_4$ | 12.65 | 0.74 |
| $C_{11}H_4N_2O_2$ | 12.79 | 1.15 |
| $C_{11}H_8N_4$ | 13.54 | 0.85 |
| $C_{11}H_{16}O_3$ | 12.26 | 1.29 |
| $C_{11}H_{20}N_2O$ | 13.01 | 0.98 |
| $C_{12}H_8N_2O$ | 13.90 | 1.09 |
| $C_{12}H_{20}O_2$ | 13.37 | 1.22 |
| $C_{12}H_{24}N_2$ | 14.11 | 0.92 |
| $C_{13}H_8O_2$ | 14.25 | 1.34 |
| $C_{13}H_{12}N_2$ | 15.00 | 1.05 |
| $C_{13}H_{24}O$ | 14.47 | 1.17 |
| $C_{14}H_{12}O$ | 15.36 | 1.30 |
| $C_{14}H_{28}$ | 15.58 | 1.13 |
| $C_{15}H_{16}$ | 16.47 | 1.27 |
| | | |
| **197** | | |
| $C_8H_{11}N_3O_3$ | 10.08 | 1.06 |
| $C_9H_{11}NO_4$ | 10.44 | 1.29 |
| $C_9H_{15}N_3O_2$ | 11.19 | 0.97 |
| $C_{10}H_{15}NO_3$ | 11.55 | 1.21 |
| $C_{10}H_{19}N_3O$ | 12.29 | 0.90 |
| $C_{11}H_7N_3O$ | 13.18 | 1.00 |
| $C_{11}H_{19}NO_2$ | 12.65 | 1.14 |
| $C_{11}H_{23}N_3$ | 13.40 | 0.83 |
| $C_{12}H_7NO_2$ | 13.54 | 1.25 |
| $C_{12}H_{11}N_3$ | 14.29 | 0.95 |
| $C_{12}H_{23}NO$ | 13.76 | 1.08 |
| $C_{13}H_{11}NO$ | 14.64 | 1.20 |
| $C_{13}H_{27}N$ | 14.86 | 1.03 |
| $C_{14}H_{15}N$ | 15.75 | 1.16 |
| | | |
| **198** | | |
| $C_8H_{10}N_2O_4$ | 9.72 | 1.23 |
| $C_8H_{14}N_4O$ | 10.47 | 0.90 |
| $C_9H_{14}N_2O_3$ | 10.83 | 1.13 |
| $C_9H_{18}N_4O$ | 11.58 | 0.82 |
| $C_{10}H_6N_4O$ | 12.47 | 0.92 |
| $C_{10}H_{14}O_4$ | 11.19 | 1.37 |

| | M + 1 | M + 2 |
|---|---|---|
| $C_{10}H_{18}N_2O_2$ | 11.94 | 1.05 |
| $C_{10}H_{22}N_4$ | 12.68 | 0.74 |
| $C_{11}H_6N_2O_2$ | 12.82 | 1.16 |
| $C_{11}H_{10}N_4$ | 13.57 | 0.85 |
| $C_{11}H_{18}O_3$ | 12.29 | 1.29 |
| $C_{11}H_{22}N_2O$ | 13.04 | 0.99 |
| $C_{12}H_6O_3$ | 13.18 | 1.40 |
| $C_{12}H_{10}N_2O$ | 13.93 | 1.10 |
| $C_{12}H_{22}O_2$ | 13.40 | 1.23 |
| $C_{12}H_{26}N_2$ | 14.15 | 0.93 |
| $C_{13}H_{10}O_2$ | 14.29 | 1.35 |
| $C_{13}H_{14}N_2$ | 15.04 | 1.05 |
| $C_{13}H_{26}O$ | 14.50 | 1.18 |
| $C_{14}H_{14}O$ | 15.39 | 1.30 |
| $C_{14}H_{30}$ | 15.61 | 1.14 |
| $C_{15}H_{18}$ | 16.50 | 1.27 |
| $C_{16}H_6$ | 17.39 | 1.42 |
| | | |
| **199** | | |
| $C_8H_{13}N_3O_3$ | 10.11 | 1.06 |
| $C_9H_{13}NO_4$ | 10.47 | 1.30 |
| $C_9H_{17}N_3O_2$ | 11.22 | 0.98 |
| $C_{10}H_5N_3O_2$ | 12.11 | 1.07 |
| $C_{10}H_{15}O_4$ | 11.20 | 1.37 |
| $C_{10}H_{17}NO_3$ | 11.58 | 1.21 |
| $C_{10}H_{21}N_3O$ | 12.33 | 0.90 |
| $C_{11}H_5NO_3$ | 12.47 | 1.31 |
| $C_{11}H_9N_3O$ | 13.21 | 1.01 |
| $C_{11}H_{21}NO_2$ | 12.68 | 1.14 |
| $C_{11}H_{25}N_3$ | 13.43 | 0.84 |
| $C_{12}H_9NO_2$ | 13.57 | 1.25 |
| $C_{12}H_{13}N_3$ | 14.32 | 0.95 |
| $C_{12}H_{25}NO$ | 13.79 | 1.08 |
| $C_{13}H_{13}NO$ | 14.68 | 1.20 |
| $C_{13}H_{29}N$ | 14.89 | 1.03 |
| $C_{14}H_{17}N$ | 15.78 | 1.16 |
| $C_{15}H_5N$ | 16.67 | 1.30 |
| | | |
| **200** | | |
| $C_8H_{12}N_2O_4$ | 9.76 | 1.23 |
| $C_8H_{16}N_4O_2$ | 10.51 | 0.90 |
| $C_9H_4N_4O_2$ | 11.39 | 0.99 |
| $C_9H_{16}N_2O_3$ | 10.86 | 1.14 |
| $C_9H_{20}N_4O$ | 11.61 | 0.82 |
| $C_{10}H_8N_4O$ | 12.50 | 0.92 |
| $C_{10}H_{16}O_4$ | 11.22 | 1.37 |
| $C_{10}H_{20}N_2O_2$ | 11.97 | 1.06 |
| $C_{10}H_{24}N_4$ | 12.72 | 0.75 |
| $C_{11}H_8N_2O_2$ | 12.86 | 1.16 |
| $C_{11}H_{12}N_4$ | 13.60 | 0.86 |

| | M + 1 | M + 2 |
|---|---|---|
| $C_{11}H_{20}O_3$ | 12.32 | 1.30 |
| $C_{11}H_{24}N_2O$ | 13.07 | 0.99 |
| $C_{12}H_8O_3$ | 13.21 | 1.40 |
| $C_{12}H_{12}N_2O$ | 13.96 | 1.10 |
| $C_{12}H_{24}O_2$ | 13.43 | 1.23 |
| $C_{12}H_{28}N_2$ | 14.18 | 0.93 |
| $C_{13}H_{12}O_2$ | 14.32 | 1.35 |
| $C_{13}H_{16}N_2$ | 15.07 | 1.06 |
| $C_{13}H_{28}O$ | 14.54 | 1.18 |
| $C_{14}H_{16}O$ | 15.42 | 1.31 |
| $C_{15}H_{20}$ | 16.53 | 1.28 |
| $C_{16}H_8$ | 17.42 | 1.42 |

# Expected Fragmentation Patterns According to Compound Type

| Compound type | Expected peaks $(m/z)^a$ | Structural unit or fragment |
|---|---|---|
| 1. Hydrocarbons | | |
|    A. Alkanes | 15, 29, 43, 57, 71 .... | $C_nH_{2n+1}{}^+$ |
|    B. Cycloalkanes | 41, 55, 69, 83 .... | $C_nH_{2n-1}{}^+$ |
| | $M-28/M-42/M-56$ ... | $-RCH{=}CH_2$ |
|    C. Alkenes | 41, 55, 69, 83 .... | $C_nH_{2n-1}{}^+$ |
|    D. Benzenoid compounds | 77 | $C_6H_5{}^+$ |
| | $\left.\begin{array}{c}91 \\ 65\end{array}\right] - HC{\equiv}CH$ | $C_7H_7{}^+$<br>$C_5H_5{}^+$ |
| | 92 | $C_7H_8{}^+$ |
| 2. Halocarbons | | |
|    A. Chlorides$^b$ | $M-36$ | $-HCl$ |
| | 91 | $C_4H_8Cl^+$ |
|    B. Bromides$^b$ | $M-80$ | $-HBr$ |
| | 135 | $C_4H_8Br^+$ |
| 3. Hydroxy compounds | | |
|    A. Alcohols$^b$ | $M-2$ | $RCH{=}O^+$ |
| | $M-3$ | $RC{\equiv}O^+$ |
| | $M-18$ | $-H_2O$ |
| | $M-33$ | $-H_2O, -CH_3$ |
| | $M-46$ | $-H_2O, -CH_2{=}CH_2$ |
|      $1°$ | 31 | $CH_2{=}OH^+$ |
|      $2°$ | 45/59/73 ... | $RCH{=}OH^+$ |
|      $3°$ | 59/73/87 ... | $R_2C{=}OH^+$ |
|    B. Phenols | $M-28$ | $-CO$ |
| | $M-29$ | $-CHO$ |
| 4. Ethers | | |
|    A. Aliphatic$^b$ | 31, 45, 59, 73 ... | $RO^+, ROCH_2{}^+$ |
| | 29, 43, 57 .... | $R^+$ |
|    B. Aromatic$^c$ | $\left.\begin{array}{l}92 + \text{mass of ring} \\ \quad\text{substituent(s), X} \\ 64 + \text{mass of ring} \\ \quad\text{substituent(s), X}\end{array}\right] - CO$ | $X{-}C_6H_4O^+$<br><br>$X{-}C_5H_4{}^+$ |
| | 76 + mass of ring<br>   substituent(s), X | $X{-}C_6H_4{}^+$ |

$^a$ Series of numbers separated by commas (i.e. 15, 29, 43 ...) represent several peaks that should be observed. Series of numbers separated by slashes (i.e. 61/75/89) indicate the possible $m/z$ values at which a single peak should be observed, depending on the mass of the attached alkyl group.

$^b$ For these aliphatic compounds, the alkyl series 15, 29, 43 ... will also be observed.

$^c$ For simple benzene derivatives, X = H; add 1 to the given $m/z$ values.

| | | |
|---|---|---|
| **5. Aldehydes** | | |
| A. Aliphatic[b] | $M - 1$ | $- H$ |
| | $M - 29$ | $- CHO$ |
| | 44/58/72 ... | $- RCH = CH_2$ |
| | | (McLafferty rearrangement) |
| B. Aromatic | $M - 1$ | $- H$ |
| | $M - 2$ | $- H, - CO$ |

| | | |
|---|---|---|
| **6. Ketones** | | |
| A. Aliphatic[b] | 58/72/86 ... | $- RCH = CH_2$ |
| | | (McLafferty rearrangement) |
| B. Aromatic | 76 + mass of ring substituent(s) | $X - C_6H_4^+$ |
| | 104 + mass of ring substituent(s) | $X - C_6H_4CO^+$ |

| | | |
|---|---|---|
| **7. Carboxylic acids and esters** | | |
| A. Acids[b] | $M - 17$ | $- OH$ |
| | $M - 45$ | $- COOH$ |
| | $M - 28/M - 42/M - 56$ ... | $- RCH = CH_2$ |
| | | (McLafferty rearrangement) |
| B. Methyl esters | $M - 31$ | $- OCH_3$ |
| | $M - 59$ | $- CO_2CH_3$ |
| | $M - 28/M - 42/M - 56$ ... | $- RCH = CH_2$ |
| | | (McLafferty rearrangement) |

| | | |
|---|---|---|
| **8. Amines** | | |
| A. Aliphatic | 30, 44, 58, 72 ... | $RCHNH_2^+$ |
| B. Aromatic | $M - 1$ | $- H$ |
| | $M - 27$ | $- HCN$ |
| | $M - 15/M - 29/M - 43$ | alkyl cleavage from N |

| | | |
|---|---|---|
| **9. Amides** | $M - 15/M - 29/M - 43$ ... | alkyl cleavage from N |
| | $M - 28/M - 42/M - 56$ ... | $- RCH = CH_2$ |
| | | (McLafferty rearrangement) |

| | | |
|---|---|---|
| **10. Nitriles** | | |
| A. Aliphatic | $M - 1$ | $- H$ |
| | $M - 28/M - 42/M - 56$ ... | $- RCH = CH_2$ |
| | | (McLafferty rearrangement) |
| B. Aromatic | $M - 1$ | $- H$ |
| | $M - 27$ | $- HCN$ |

| | | |
|---|---|---|
| **11. Nitro compounds (aromatic)** | | |
| | $M - 30$ | $- NO$ (with rearrangement) |
| | $M - 46$ | $- NO_2$ |
| | $M - 58$ | $- NO, - CO$ |
| | $M - 72$ | $- NO_2, - HC \equiv CH$ |

| | | |
|---|---|---|
| **12. Thiols and sulfides[b]** | | |
| A. Thiols | $M - 34$ | $- H_2S$ |
| 1° | 47 | $CH_2 = SH^+$ |
| 2° | 61/75/89 ... | $RCH = SH^+$ |
| 3° | 75/89/103 ... | $R_2C = SH^+$ |
| B. Sulfides | 47/61/75 ... | $RS^+$ |

# Index

**169**